AI and Ethics
A computational perspective

Online at: https://doi.org/10.1088/978-0-7503-6116-3

IOP Series in Next Generation Computing

Series editors
Prateek Agrawal
University of Klagenfurt, Austria and Lovely Professional University, India

Anand Sharma
Mody University of Science and Technology, India

Vishu Madaan
Lovely Professional University, India

About the series
The motivation for this series is to develop a trusted library on advanced computational methods, technologies, and their applications.

This series focuses on the latest developments in next generation computing, and in particular on the synergy between computer science and other disciplines. Books in the series will explore new developments in various disciplines that are relevant for computational perspective including foundations, systems, innovative applications, and other research contributions related to the overall design of computational tools, models, and algorithms that are relevant for the respective domain. It encompasses research and development in artificial intelligence, machine learning, block chain technology, quantum cryptography, quantum computing, nanoscience, bioscience-based sensors, IoT applications, nature inspired algorithms, computer vision, bioinformatics, etc, and their applications in the areas of science, engineering, business, and the social sciences. It covers a broad spectrum of applications in the community, including those in industry, government, and academia.

The aim of the series is to provide an opportunity for prospective researchers and experts to publish works based on next generation computing and its diverse applications. It also provides a data-sharing platform that will bring together international researchers, professionals, and academics. This series brings together thought leaders, researchers, industry practitioners and potential users of different disciplines to develop new trends and opportunities, exchange ideas and practices related to advanced computational methods, and promote interdisciplinary knowledge.

A full list of titles published in this series can be found here: https://iopscience.iop.org/bookListInfo/iop-series-in-next-generation-computing.

AI and Ethics
A computational perspective

Animesh Mukherjee
*Department of Computer Science and Engineering, Indian Institute of
Technology Kharagpur (India), Kharagpur, India*

IOP Publishing, Bristol, UK

© IOP Publishing Ltd 2023

All rights reserved. No part of this publication may be reproduced, stored in a retrieval system or transmitted in any form or by any means, electronic, mechanical, photocopying, recording or otherwise, without the prior permission of the publisher, or as expressly permitted by law or under terms agreed with the appropriate rights organization. Multiple copying is permitted in accordance with the terms of licences issued by the Copyright Licensing Agency, the Copyright Clearance Centre and other reproduction rights organizations.

Permission to make use of IOP Publishing content other than as set out above may be sought at permissions@ioppublishing.org.

Animesh Mukherjee has asserted his right to be identified as the author of this work in accordance with sections 77 and 78 of the Copyright, Designs and Patents Act 1988.

ISBN   978-0-7503-6116-3 (ebook)
ISBN   978-0-7503-6112-5 (print)
ISBN   978-0-7503-6113-2 (myPrint)
ISBN   978-0-7503-6115-6 (mobi)

DOI   10.1088/978-0-7503-6116-3

Version: 20230701

IOP ebooks

British Library Cataloguing-in-Publication Data: A catalogue record for this book is available from the British Library.

Published by IOP Publishing, wholly owned by The Institute of Physics, London

IOP Publishing, No.2 The Distillery, Glassfields, Avon Street, Bristol, BS2 0GR, UK

US Office: IOP Publishing, Inc., 190 North Independence Mall West, Suite 601, Philadelphia, PA 19106, USA

My physical disability that taught me how empathy is different from sympathy

My beloved wife, Soumita, for the umpteen cups of tea

My loveliest daughter, Anugyaa, for lending me the questioning eye of a child

Contents

Preface	xi
Acknowledgements	xiii
Author biography	xv

1 Preamble — 1-1

1.1 What is ethics? — 1-1
    1.1.1 Reasoning in ethics — 1-1
    1.1.2 Ethical dilemma — 1-2
1.2 What is AI? — 1-3
    1.2.1 The first narrow down: machine learning — 1-4
    1.2.2 The second narrow down: deep learning — 1-5
1.3 What is AI and ethics? — 1-9
    1.3.1 AI and human rights — 1-9
    1.3.2 AI and finance — 1-11
    1.3.3 AI and the law — 1-11
    1.3.4 Bias and fairness — 1-12
1.4 What is a computational perspective? — 1-13
    References — 1-14

2 Discrimination, bias, and fairness — 2-1

2.1 Chapter foreword — 2-1
2.2 Sensitive attributes — 2-1
2.3 Notions of discrimination — 2-2
    2.3.1 Disparate treatment — 2-3
    2.3.2 Disparate impact — 2-3
    2.3.3 Disparate mistreatment — 2-4
2.4 Types of bias — 2-4
    2.4.1 Confirmation bias — 2-4
    2.4.2 Population bias — 2-4
    2.4.3 Sampling/selection bias — 2-4
    2.4.4 Interaction bias — 2-5
    2.4.5 Decline bias — 2-5
    2.4.6 The Dunning–Kruger effect — 2-5
    2.4.7 Simpson's paradox — 2-5

	2.4.8 Hindsight bias	2-5
	2.4.9 In-group bias	2-6
	2.4.10 Emergent bias	2-6
2.5	Operationalizing fairness	2-6
	2.5.1 Demographic/statistical parity	2-6
	2.5.2 Equal opportunity	2-6
	2.5.3 Predictive parity	2-7
	2.5.4 Equalized odds	2-7
	2.5.5 Disparate mistreatment	2-7
	2.5.6 Fairness through (un)awareness	2-7
2.6	Relationships between fairness metrics	2-7
	2.6.1 Connection between statistical parity, predictive parity, and equalized odds	2-8
	2.6.2 Statistical and predictive parity	2-8
	2.6.3 Statistical parity and equalized odds	2-8
	2.6.4 Predictive parity and equalized odds	2-9
2.7	Individual versus group fairness	2-9
	Practice problems	2-10
	References	2-11

3 Algorithmic decision making — 3-1

3.1	Chapter foreword	3-1
3.2	The ProPublica report	3-1
	3.2.1 Data gathered by the authors	3-1
	3.2.2 Defining recidivism	3-2
	3.2.3 Primary observations	3-2
3.3	Gender shades	3-4
	3.3.1 Biases in existing gender classification benchmark datasets	3-4
	3.3.2 Construction of a new benchmark dataset	3-5
	3.3.3 Biases in gender detection	3-5
3.4	Fair ML	3-5
	3.4.1 Fair classification	3-5
	3.4.2 Fair regression	3-9
	3.4.3 Fair clustering	3-10
	3.4.4 Fair embedding	3-12
	Practice problems	3-16
	References	3-18

4 Content governance — 4-1
- 4.1 Chapter foreword — 4-1
- 4.2 Content dissemination — 4-2
  - 4.2.1 Discrimination in Facebook ad delivery — 4-2
  - 4.2.2 Unfairness against third-party products and sellers in Amazon — 4-4
- 4.3 Content moderation — 4-7
  - 4.3.1 Fake news — 4-8
  - 4.3.2 Hate speech — 4-14
  - 4.3.3 Media bias — 4-26
  - Practice problems — 4-30
  - References — 4-34

5 Interpretable versus explainable models — 5-1
- 5.1 Chapter foreword — 5-1
- 5.2 What is interpretability? — 5-1
- 5.3 What is explainability? — 5-2
- 5.4 Operationalizing interpretability — 5-3
  - 5.4.1 Partial dependence plot — 5-3
  - 5.4.2 Functional decomposition — 5-4
  - 5.4.3 Statistical regression — 5-5
  - 5.4.4 Individual conditional expectation — 5-5
- 5.5 Operationalizing explainability — 5-6
  - 5.5.1 Local surrogate — 5-6
  - 5.5.2 Anchors — 5-7
  - 5.5.3 Shapley values — 5-8
  - 5.5.4 SHAP — 5-10
  - 5.5.5 Counterfactual explanations — 5-11
  - Practice problems — 5-14
  - References — 5-14

6 Newly emerging paradigms — 6-1
- 6.1 Chapter foreword — 6-1
- 6.2 Ethical use of AI for law enforcement — 6-1
  - 6.2.1 Ethical concerns in predictive policing — 6-1
  - 6.2.2 Violation of fundamental human rights — 6-2
  - 6.2.3 Possible alternatives — 6-2

6.3	Data protection and user privacy	6-3
	6.3.1 Data protection	6-5
	6.3.2 Cybersecurity	6-5
6.4	Moral machines	6-6
	6.4.1 Overall observations	6-7
	6.4.2 Clusters based on culture	6-7
	6.4.3 Collectivist versus individualistic cultures	6-8
	6.4.4 Underdeveloped versus developed countries	6-8
	6.4.5 Economic differences	6-8
6.5	Singularity and ethics	6-8
	6.5.1 Is singularity imminent?	6-9
	6.5.2 What are the existential risks from singularity?	6-10
	6.5.3 Is it possible to declaw the singularity?	6-10
	References	6-11

Preface

AI is more than 65 years old now and has seen many ups and downs on its journey. Never before has it changed human lives so quickly and at such a scale as it is doing now. This burgeoning success is often referred to by experts as the AI boom. Nevertheless, there are no roses without thorns! Such massive and complex AI systems could unknowingly cause certain groups of people to come to harm or could even be knowingly abused to bring harm to them. Classic examples include the *Gender Shades* project by Joy Buolamwini at MIT Media Lab or the *Machine Bias* project by ProPublica.

Many of these AI systems are typically developed by engineers with Computer Science majors and awareness about such harms could make them morally and ethically alert when they build such software. One of the best ways to possibly spark this awareness in them is to start early while they are still in their undergraduate years and expose them to a semester-long course on how to design ethically responsible AI software. In 2019, I made an effort to propose such a course under the name of *AI and Ethics* in my Department. Immediately, there was an uproar in the Department that a course on 'Ethics' could be appropriate for CS major students. The main issue was the idea that a combination of experts from the law and the humanities departments be allowed to manage such courses. The course and this book would have been doomed then and there without a brilliant idea from one of my colleagues who was very positive about the course and its outcomes. His suggestion was to reach out to software industry experts asking for the timely relevance of such a course. I received an overwhelming response from the industry citing how relevant such ethical knowledge is in developing large AI systems (and in fact, any large software for that matter). A summary of these suggestions was that not only such a course is needed but that it needs to be taught from a computational perspective. This support allowed me to actually float the course in Spring 2020 for the first time. The number of registrations for the course was again overwhelming and was opted for by students from all engineering disciplines with a large majority coming from electrical sciences. As we started the course, the students soon felt that a textbook was needed for the course which; so far this still did not exist. The course ran successfully for the next two Springs (Spring 2021 and Spring 2022) and my inbox was flooded with emails like the one reproduced below.

Dear Sir

I am XXXX, a student from the AI and Ethics course you took this spring. I just wanted to write this email to thank you for the wonderful way in which you taught this class.

Admittedly, I was unsure about taking this course at first because I had heard it might be too much effort, but this was probably one of the best courses I have taken in KGP so far.

> The efforts I put in were all worth it because ultimately I was left with a satisfactory feeling about what I was doing and working and the term project felt like we were doing something truly impactful. I really wished we could have had this class offline as it would have given rise to many more conversations, but nevertheless, I truly enjoyed studying for this course. More importantly, it introduced me to a whole new side of Machine Learning and now has me questioning the ethical aspect of any new algorithm I come across.
>
> Once again, thank you for this course :)

While this was really heartening, the absence of a relevant textbook was constantly felt by the students and teaching assistants, and also by myself as the instructor.

Toward the end of 2021, I was invited by the Institite of Physics to write a book on some of the recent issues in information retrieval. It was then that some of my past students from the AI and Ethics course persuaded me and, in fact, succeeded in having me sign the contract with IOP to write this textbook. The hard workl then actually began. Thanks to the course content that I had developed together with my teaching assistants, the organization of the chapters was to much challenging! The amount of literature that I got to read and understand while compiling the individual chapters was amazing. I hope my readers will also benef from this knowledge. Among the many challenges that came up while writing this book and the one that I most enjoyed was creating relevant examples in different places to motivate a theory or explain a concept. Overall the exercise was a great mix of learning and fun which has resulted in this textbook which as I see as one of the treasures of my lifetime.

Acknowledgements

No major initiative in life is ever successful unless there is a generous flow of support and cooperation from the people surrounding us. IThe writing of this book has not an exception. Therefore, it is imperative that due gratitude be expressed to all of them.

The first person that I need to profusely thank is my dear colleague who protested against teaching the AI and Ethics course in our Department. If I had not been challenged by them and not had to face the follow up crisis I would never have had the conviction to push the course through and write this book. Looking back I feel that this was the golden push that took me forward!

Abhisek and **Siddharth**, two of my beloved PhD students have made painstaking efforts to thoroughly proofread the contents the book, improve it in places and enrich it with pertinent citations and their own ideas. I have intentionally emboldened their names to demonstrate their exemplary contribution to the drafting of this book and my heartfelt gratitude toward them. Without their efforts this book would have never been completed.

As I mentioned in the preface, this book is a direct product of my course and no course can be realized without a remarkable set of teaching assistants. I need to thank Kalyani Roy, and another group of my fantastic PhD students—Binny Mathew, Punyajoy Saha, Souvic Chakraborty and Sayantan Adak who have all been very proactive in shaping this course, bringing in their own bits of knowledge and thus indirectly enriching the contents of this book. Similarly, the students taking up this course have also been very instrumental in bringing new ideas to enhance the course content and, subsequently, this book.

I also need to thank many colleagues from whom I have learned the importance of this field. In this context, I have been fortunate to have had the opportunity to collaborate with Krishna P Gummadi from MPI, SWS. Krishna spearheaded the development of the area of fairness in machine learning right from the beginning. Not only have I learned the art of doing research in this area from him but also have developed a very friendly relationship with him and many of our over-the-coffee discussions have become motivating examples in this book. My long-time friend and collaborator Chris Biemann from the University of Hamburg, educated me about the ethical issues in natural language processing and what could be some of the possible ways to tackle these. Such discussions have materialized into the substantial content of this book. My collaborations with Kiran Garimella from Rutgers University have taught me about the issues in content governance and especially content moderation at a more microscopic level and this lesson has appeared as discussions in various parts of the book.

People say 'home sweet home', I say 'family sweet family'. Without the warmth, love, and compassion that I received from my beloved wife and my charming daughter this book would have just remained a dream. Not only their love but also their ethical understanding helped me to appreciate its importance in daily life—just as they say 'charity begins at home'. Some anecdotal examples of such learnings are here.

When my daughter was five, I was on a sabbatical at the University of Hamburg. We dropped her in a kindergarten where she inculcated two important ethical values—wasting food is a crime and littering public places is as bad as being antisocial. Since then she firmly upholds these moral values and is the first in our family to stop us from committing the aforementioned crimes. My wife is the first person who showed me that people with physical disabilities or those who are socially or financially challenged need empathy and not sympathy. Her elevated thoughts have helped me to go beyond my physical limitations and take up the very challenging task of writing a book without facing practically any hazard. Kudos to both of these wonderful people! My parents Mrs Seema Mukherjee and Mr Kumaresh Mukherjee have always led a very simple life and I am indebted to them for inculcating this simplicity in me. This simple viewpoint toward life has also helped me in simplifying things presented in various parts of this book which I feel should be beneficial for the reader.

Author biography

Animesh Mukherjee

Presently, Animesh Mukherjee is a Full Professor in the Department of Computer Science and Engineering, Indian Institute of Technology, Kharagpur. He is also a Distinguished Member of ACM. His main research interests center around content governance which includes (i) content moderation (harmful content analysis, detection, and mitigation), (ii) content dissemination (fairness issues in e-commerce platforms and interfaced systems like facial recognition, automatic speech recognition systems, etc), and (iii) content maintenance (quality analysis and improvement of encyclopedias like Wikipedia and large software systems like Ubuntu releases). He regularly publishes in all top CS conferences including AAAI, IJCAI, ACL, NAACL, EMNLP, The Web Conference, CSCW, etc. He has received many notable awards and fellowships including the Facebook ethics for AI research award, India, Google course award for the course AI and Ethics, IBM faculty award, Humboldt Fellowship for Experienced Researchers, Simons Associateship, ICTP, to name a few.

IOP Publishing

AI and Ethics
A computational perspective
Animesh Mukherjee

Chapter 1

Preamble

1.1 What is ethics?

Traditionally, *ethics* has been classified as a branch of philosophy that is supposed to present a systematic understanding of the differences between right and wrong, the just and the unjust and, the good and the bad in relationships, to the well-being of the mortals gifted with the unique ability of sentience. Usually, ethics is believed to be always in 'action'; hence is more suitably termed as *doing ethics*. *Morals*, as opposed to ethics, are certain beliefs, actions, and behaviors that are formulated based on ways of doing ethics aka *ethical standards*. However, these two terms have been used interchangeably in most of the earlier literature and we shall follow the same convention in this book unless otherwise specified.

1.1.1 Reasoning in ethics

While doing ethics, it is of utmost importance to justify one's thoughts and assertions through sound arguments and solid reasoning. Although, it may be argued that the design of ethical standards is subjective, it should always be possible to justify these standards using theoretically grounded or well-formed arguments. While the emotions and feelings of a society can always play a part in the design of its ethical standards, they should not overtake the role of logic and reasoning. In cases when emotions and feelings overtake, there is a potential for harmful biases slipping into the design of these standards.

Reasoning is the process of receiving information, comparing it with what is already known and then coming to a conclusion. There are primarily two types of reasoning—*deductive* reasoning and *inductive* reasoning. Deductive reasoning is the process of deducing logically valid conclusions from a set of general statements. Inductive reasoning, on the other hand, is the process of arriving at a general conclusion from various specific elements of information.

Ethical reasoning pertains to reasoning based on ethical standards and can be both deductive and inductive in nature. An example of a deductive ethical reasoning could be as follows (**P** is a proposition and **C** is a conclusion).

P1: it is ethically incorrect to not shortlist a competent person for a job position based on ethnicity.
P2: it is an act of a black male being not shortlisted for a job position.
P3: the black male is competent for the job position.
C: this act is ethically incorrect.

On the other hand, an inductive ethical reasoning would be of the following form.

P1: any act of racial genocide is ethically wrong.
P2: the extermination of the Rohingya Muslims by the Buddhist nationalists and the military of Myanmar was an act of racial genocide.
C: the extermination of the Rohingya Muslims by the Buddhist nationalists and the military of Myanmar was ethically wrong.

1.1.2 Ethical dilemma

Moral or ethical dilemmas are paradoxical scenarios where a person has to choose between two or more conflicting moral situations neither of which overrides the other. One famous example of such a moral dilemma is the 'trolley predicament'. The situations and the dilemma are presented below.

There is a trolley moving on the railway track as shown in figure 1.1. At some distance from the trolley there are five people tied to the track unable to move or escape. The trolley is directly headed toward them. At some distance, there is Binny standing with the control of the switch/lever. If Binny chooses to pull the lever, the trolley will deflect to a side track. But there is one person on the side track trying to cross the track and they would be surely hit if the trolley is deflected. Now Binny has two (and only two) options (**S1** and **S2**).
  S1: Binny does nothing, in which case the five people tied to the main track will be killed.
  S2: Binny pulls the lever, in which case the trolley is deflected to the side track and kills the person crossing it.
Dilemma: what is the morally correct thing to do for Binny?

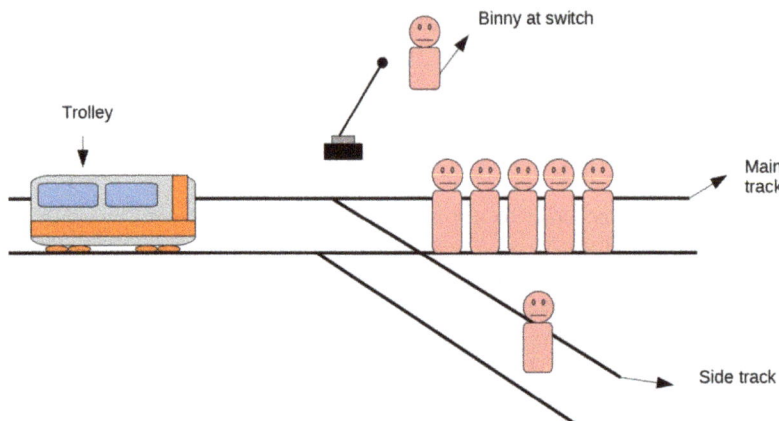

Figure 1.1. The trolley predicament. Binny has two (and only two) choices—(a) do nothing and allow the trolley to run on the main track and kill the five people, or, (b) pull the switch/lever to deflect the train to the side track, thereby, killing the person trying to cross the track.

There have been many studies on these types of moral decision problems and a clear solution to the problem is still not in place. The only weak conclusion that researchers have reached so far is that moral reasoning can sometimes be ineffective in either the induction or deduction of judgements based on well-established ethical standards (e.g., killing in all forms is morally wrong). This means that it might not be always possible to operationalize moral reasoning through a single formal model.

1.2 What is AI?

Artificial intelligence (AI) is the science and engineering of making computers mimic the decision-making and problem-solving skills of humans. Russel and Norvig [1] in their popular textbook describe AI systems as those that think and act like humans (humanistic approach) or those that think and act rationally (idealistic approach).

Serious research on non-human intelligence has at least 80 years of solid history. It started with Alan Turing's work on how machines can solve deductive problems by operating on zeroes and ones. This was a huge breakthrough as it gave the *machine the power of reasoning*. The promise of this research was soon felt to be far reaching and expectations mounted across the world. Money started pouring in from all corners in the quest to develop various captivating applications. However, it is well-known that soaring claims around a new technology can often backfire. AI was no exception. Soon it was realized that the expectations were too far ahead of the reality. Unfortunately, the shocks of such failures are carried long in the memory and funding soon dries up. In the history of AI such periods have appeared at least twice (1974–80 and 1987–93) and are popularly known as *AI winters*. The first of these marks the failure of machine translation systems during the Cold War. It was soon felt that phrase sense disambiguation (which is typically done by humans using common sense knowledge) was a reasonably hard problem and is one of the many things that need to be solved to perfect machine translation.

The second winter was caused by a split in the AI community. One of these groups were the proponents of building large and commercial expert systems developed on the premises of symbolic logic. However, these systems were soon found to be too hard to maintain and lost traction. The other part of the community went on to develop the connectionist approach (based on artificial neurons and their interactions) which was much more flexible but could only be developed theoretically due to limits on the processing power of the computers. Among many other things, these AI winters had a narrowing effect on the overall scope of AI. Two such consecutive narrowing downs helped in thawing the 'ice age' of AI.

1.2.1 The first narrow down: machine learning

Machine learning is a sub-branch of AI that uses data and algorithms to perform reasoning in a way that mimics human reasoning. The accuracy in the reasoning is proportional to the amount of the data that the algorithm observes (or learns from). Algorithms are developed based on various statistical methods trained to make classifications of the input data. Usually such methods require the data to be well-structured. Human intervention is necessary to determine the features that can characterize the differences between the classes of input data. Such features are then used to train a statistical model and, thereby, make inferences on unseen data. There are mainly two broad paradigms of machine learning problems.

Supervised learning. In the supervised learning setup the main goal is to estimate an *accurate* predictor function $f(x)$ which is usually known as the hypothesis. *Learning* is the process of optimizing this function given some input data x (e.g., temperature, air pressure, wind velocities, wind directions, and precipitation amounts). The task is to predict some interesting outcome variable for $f(x)$ (e.g., the amount of rainfall) as accurately as possible. In general, x has multiple attributes, e.g., temperature (x_1), air pressure (x_2), wind velocities (x_3), wind directions (x_4), and precipitation amounts (x_5). Identifying which set of inputs/features would be effective for the prediction is a design choice and needs human intervention. Without any loss of generality let us assume the simplest single attribute case where the predictor would take the form

$$f(x) = a + bx \tag{1.1}$$

The challenge here is to estimate the values of a and b that makes $f(x)$ accurate. This estimation is done using a number of training examples of the form (x_{train}, y) which means that the correct output for x_{train} is y and is known in advance. Each time we make a prediction using $f(x)$ and check how far it deviates from y. In the optimization step, one attempts to adjust the values of a and b to reduce the deviation thus making the predictor $f(x)$ 'less wrong' as it is able to see more training examples. This continues for a while until the values of a and b finally converge and the deviation cannot be minimized any further. At this point the predictor is deemed suitable to make predictions for some real-world application.

The deviation is usually measured in the form of a *loss* function that iteratively improves the values of a and b. If we denote $L(a, b)$ as the loss function then one of the many ways to define this function could be to use the least squares regression as follows

$$L(a, b) = \frac{1}{n}\sum_{i=1}^{n}(f(x_{\text{train},i}) - y_i)^2 \tag{1.2}$$

Thus we need to obtain the values of a and b in our predictor function $f(x)$ such that $L(a, b)$ is as small as possible. This is solved using some simple calculus. Generally, the gradient or the derivatives of L with respect to a and b are computed as follows (getting free of constants)

$$\frac{\partial L}{\partial a} \equiv \frac{\partial}{\partial a}\sum_{i=1}^{n}(f(x_{\text{train},i}) - y_i)^2 = 2\sum_{i=1}^{n}(f(x_{\text{train},i}) - y_i)\frac{\partial f(x_{\text{train},i})}{\partial a} \tag{1.3}$$

Similarly,

$$\frac{\partial L}{\partial b} \equiv 2\sum_{i=1}^{n}(f(x_{\text{train},i}) - y_i)\frac{\partial f(x_{\text{train},i})}{\partial b} \tag{1.4}$$

Now the task is to adjust $\frac{\partial f(x_{\text{train},i})}{\partial a}$ and $\frac{\partial f(x_{\text{train},i})}{\partial b}$ so that it can take us to the 'valley' of the loss function. Possibly the addition of a little value to a and subtraction of a little value from b can make the loss function go down. This iterative process of adjusting the gradients and tuning the values of a and b is called *gradient descent*. This basic calculus can be now applied to a variety of problem settings such as classification, regression, anomaly detection, etc.

Unsupervised learning. Unsupervised learning is usually used to discover latent relationships in the data. There is no labeled data in this case. Instead the algorithms are designed to identify correlations among data points to obtain tightly connected groups/communities/clusters. Some of the very well-known clustering algorithms are k-means [2], k-medoids [3], x-means [4], DBSCAN [5] etc. This family of learning algorithms also include finding principal components [6, 7], singular values [8], factor analysis [9], random projections [10, 11] etc.

1.2.2 The second narrow down: deep learning

In recent years the resurgence of AI can be mostly attributed to a sub-area of machine learning called 'deep learning'. This is based on (artificial) neural networks (ANNs) which have the capability to learn by example and do not need hand-crafted features for this. The basic architecture is thought to have a resemblance to the way our brain functions. A correspondence of the brain neuron with that of a basic unit of an ANN is shown in figure 1.2.

While the concept of ANNs were developed in the early days of AI, its true potential could only be leveraged in the last few years due to abundance of data and the cheap computational power in the form of general purpose computing on graphics processing units (GPGPUs). These techniques have enabled significant progress in various fields of computer science including image processing, computer

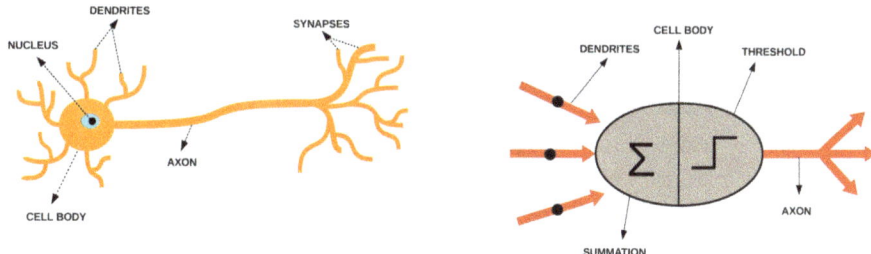

Figure 1.2. The figure on the left shows the structure of a neuron while that on the right shows the basic unit of an ANN.

vision, and natural language processing among many others. There are various architectures available but the fundamental ones can be categorized as follows

- Multilayer perceptrons which have existed from the very beginning,
- Sequence models like recurrent neural networks and long term short memory (LSTM) which are primarily used for processing sequential data like natural language text, and
- Convolution based models like convolution neural networks (CNNs) that are particularly adapted for image processing and vision problems.

Artificial neuron. An artificial neuron is defined as a function f of the input $x = [x_1, x_2, \ldots, x_d]$ which is weighted by a vector $w = [w_1, w_2, \ldots, w_d]$ plus a bias factor b and passed through an activation function ϕ thus taking the following form

$$f(x) = \phi(\langle x, w \rangle + b) \tag{1.5}$$

There can be multiple choices for ϕ; some of these include
 (i) the sigmoid function: $\phi(x) = (1 + \exp(-x))^{-1}$,
 (ii) the rectified linear unit (ReLU) function: $\phi(x) = \max(0, x)$,
 (iii) the hyperbolic tangent function $\phi(x) = \tanh(x)$ or
 (iv) simply the identity function $\phi(x) = x$.

A pictorial description of an artificial neuron is shown in figure 1.3.

Multilayer perceptron. A multilayer perceptron (aka artificial neural network) is a collection of artificial neurons stacked as an input layer followed by one or more hidden layers and a final output layer. A pictorial representation of a multilayer perceptron with a single hidden layer is shown in figure 1.4.

Generally, the output of one layer constitutes the input of the next layer. A multilayer perceptron with L hidden layers can be mathematically written as follows

$$H^{(0)}(x) = x \tag{1.6}$$

$$H^{(l)}(x) = \phi(b^{(l)} + W^{(l)}H^{(l-1)}(x)), \forall l \in L \tag{1.7}$$

$$H^{(L+1)}(x) = \psi(b^{(L+1)} + W^{(L+1)}H^{(L)}(x)) \tag{1.8}$$

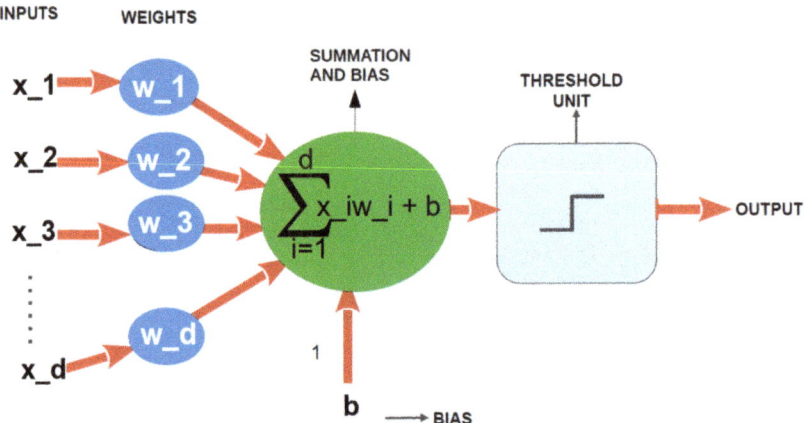

Figure 1.3. An artificial neuron.

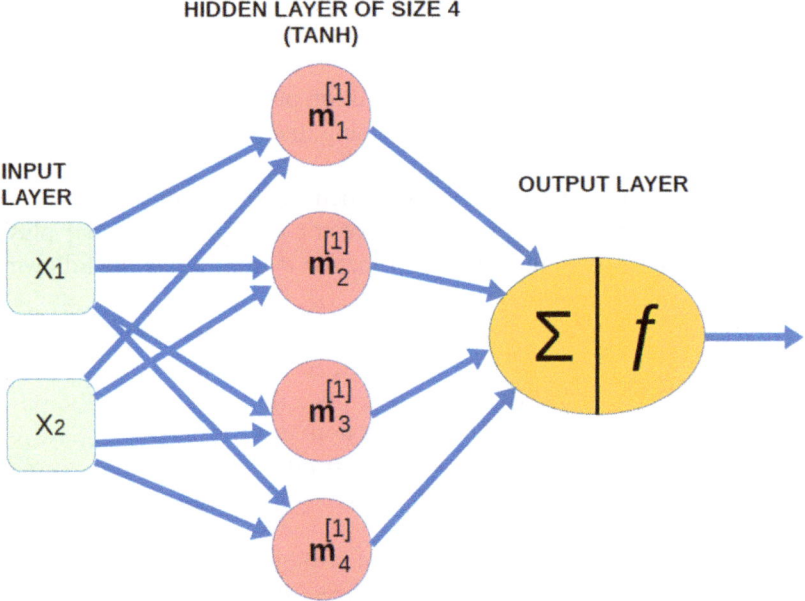

Figure 1.4. Example of a multilayer perceptron with a single hidden layer.

Here $H^{(0)}(x)$ is the input layer, $H^{(l)}(x)$ is the lth hidden layer and $H^{(L+1)}(x)$ is the output layer. $W^{(l)}$ is a matrix with the number of rows equivalent to the number of neurons in layer l and number of columns equivalent to the number of neurons in layer $l-1$. ϕ and ψ are the hidden and output layer activation functions respectively which may be the same or different based on the type of application.

Once the neurons in each layer and the number of hidden layers are fixed, the weights w and the bias factors b need to be estimated. This can be done using a set of training samples and minimizing a loss function using the same gradient descent approach described earlier.

Convolutional neural networks. CNNs are composed of several types of neuron layers—convolutional layers, pooling layers, and fully connected layers. These networks can directly work on matrices or tensors representing the RGB composition in images.

- *Convolutional layer*: in general the convolution between two function f and g is given by the following equation.

$$(f * g)[n] = \sum_m f[m]g[n+m] \tag{1.9}$$

In the case of 2D inputs such as images the function can be written as follows

$$(A * I)[i, j] = \sum_{m,n} A[m, n]I[i+n, j+m] \tag{1.10}$$

Here A is the convolution kernel and I the 2D input image. If one assumes that the kernel A is of size $a \times a$, and n is a $a \times a$ image patch then the activation is usually obtained by sliding the $a \times a$ window and computing

$$z(n) = \phi(A * n + b) \tag{1.11}$$

where ϕ is the activation function and b is a bias. Using this function, the CNN layers learn the most effective kernels (aka filters) for a given task like image classification.
- *Pooling layer*: the pooling layers are used for dimension reduction. This is done by taking the average or the maximum on the patches of image which are respectively known as *average-pooling* and *max-pooling*.
- *Fully connected layer*: following a series of convolution and pooling layers comes a set of fully connected layers. In this phase the tensor is finally converted into a vector to perform the classification.

Recurrent neural network. Recurrent neural networks (RNNs) are used to make inferences on time series data and are mostpopularly used for handling text sequences. The main idea here is to associate a notion of time with the hidden layers. The way to represent this would be to imagine that an RNN consists of multiple copies of the same network each passing information to its successor. One of the most basic forms of RNN is where a multi-layer perceptron with one unit layer is looped back on itself. If we assume that $x(t)$ is the input, $\hat{o}(t)$ the output, and $\hat{H}(t)$ the hidden layer at time t then the k component output of the model can be written as

$$\hat{o}^k(t) = \sum_{i=1}^{I} W_i^k \hat{H}_i(t) + b^k \tag{1.12}$$

$$\hat{H}_i(t) = \psi\left(\sum_{j=1}^{J} w_{i,j} x_j(t) + \sum_{l=1}^{I} \tilde{w}_{i,l} \hat{H}_l(t-1) + b_i \right) \tag{1.13}$$

where ψ is the activation function.

Recent trends. In recent times there has been a great deal of development in both CNN and RNN architectures. **AlexNet** [12] is one of the first complex CNN architectures; it won the ImageNet competition in 2012 where the task was to classify one million color images in 1000 classes. In 2014, the competition was won by **GoogleNet** [13] which had *inception* layers apart from convolution and pooling layers. The inception layer builds on the idea of a network within a network and more details on this can be found here [14]. The most recent innovation in this line of research is **ResNet** [15] where the input of a layer is connected to its output. In similar spirit, the RNN architecture have been improved to handle longer sequences. **Long short term memory (LSTM)** [16] and **gated recurrent units (GRUs)** [17] are some examples of such improvements. Currently, transformer models like **bidirectional encoder representations from transformers (BERT)** [18] and its variants have become immensely popular for text processing.

1.3 What is AI and ethics?

The remarkable rise in AI and robotic systems have led to a *range of ethical questions*. The impacts can be manifold, including but not limited to, social, psychological, environmental, legal, financial, privacy, and security. The massive prevalence of AI can have deep repercussions on human society and, in particular, the rights of individuals to a private and dignified life. If, in the long run, AI systems are able to infer one's thoughts and beliefs then it might be possible to manipulate them easily. Strategists and power leaders can use this to their benefit to persuade and even change the beliefs. Further this might lead to state leaders being able to infer someone's negative emotions toward the activities of the state which in certain countries could even result in capital punishment. Some of the direct harms that can be foreseen immediately are as follows.

1.3.1 AI and human rights

The rise of AI can have very serious implications on human rights. Signs of such threats to the human society are already in place and are expected to only grow stronger over the next few decades. Next, we outline some of these threats to further motivate the necessity of the rest of the book.

AI the spy. The use of AI to monitor and spy on every individual human activity is ever-increasing. An individual in the modern society is under constant surveillance of hundreds of interconnected cameras and other mobile/wearable devices that are not only able to monitor the location of one's home, work, or worship but can also spy on them in intimate places where privacy is of utmost importance (e.g., bathrooms, bedrooms, and changing rooms). There have been attempts to use the data collected by Amazon Alexa [19] or even a pacemaker [20] as evidence against an individual. Facial recognition systems (FRSs) are being routinely used to identify faces that are required for various authentication purposes. In their classic paper [21], Buolamwini and Gebru showed how FRSs perform poorly on certain intersectional groups always adversely affecting people of color. In a similar vein much of this software has not only been deployed to identify individuals but also to detect their moods and states of mind in schools and re-education camps [22]. Such technology can then be easily used to punish non-attentive students in schools and prisoners in re-education camps. AI may also be used to ascertain potential troublemakers and acts of recidivism. In a classic study by ProPublica on such a commercial software COMPAS, the authors showed how such systems severely discriminate against people of specific ethnicity and color [23].

Freedom of expression. Freedom of speech and expression is one of the fundamental rights in most democratic nations. This can be severely affected by the evolving AI algorithms especially because many big techs are now regularly using these algorithms for the purpose of *platform governance* [24, 25]. For instance, hate speech detection [26], misinformation detection [25, 27, 28], extremism detection [29, 30], and many other such online activities are now being routinely carried out using AI algorithms. Sentiment analysis [25, 31] tools are also being used extensively to identify the tone and emotion of the posts of the users and decisions on content removal are being taken thereof. This could be very dangerous since the current AI algorithms are far too naïve to actually understand human tones and emotions correctly. The problem is more critical when certain big techs promote such automatic censorship based on the instructions of the government of a nation. This has the potential to massively curtail the voice of the mass by encouraging self-censorship.

AI and democracy. In the past, political candidates had been mostly using small-scale surveys and word-of-mouth interactions to judge the voting inclination of individuals in a particular place. However, with the advent of social media, the politicians now have access to an unprecedented amount of data about users from which it is possible to guess their political inclinations [32] using AI algorithms. This might help the politicians to gain insights about the perception of the voters and accordingly steer the efforts of their party. However, this has its own caveats. The number of examples that (mis)uses AI technology to manipulate citizens in recent elections is abundant. For instance, there are reports of an extensive use of bots on the Twitter platform before the 2016 US Presidential elections [33]. Many of these were known to have been able to bias the content being viewed on social media thus creating a false impression of the support base for a party. Bots have also been used to influence the Brexit vote in the UK [34, 35] and to popularize the hashtag #MacronLeaks on social media before the French presidential election in 2017.

These bots are known to share a huge volume of automated content on social media most of which are fake, harmful, or spam. Some of these bots have proven to be very effective in amplifying the marginal opinions into majority. In addition to exerting collective influence, AI has also been weaponized to target individuals to manipulate their voting viewpoints. For example, before the 2016 US presidential elections Cambridge Analytica got access to the individual Facebook profiles of more than 50 million users which was used to target them with advertisements to psychologically manipulate their voting decisions [36]. A true democracy should be built on fair and free elections without any influence or manipulation in place and AI technology has presented us with enough evidence of undermining this in the recent past.

1.3.2 AI and finance

The finance and banking sectors have been strongly influenced by data science and AI for a very long time. It is well accepted that algorithmic trading can garner profits at a scale and speed that is unimaginable by a human trader. Hence a large number of companies have invested huge amounts of money to pump in more and more intelligence to the algorithms that make their trading decisions. Here again, autonomous trading agents can be designed to play malicious roles to make the trade market unstable which might have huge financial repercussions and seriously harm many small players. Even if these agents are not malicious, it is not understood how they would react to sudden changes in the market conditions as the algorithms are often not explainable.

In the banking sector one of the major evils of AI is seen in credit scoring [37]. Algorithms are routinely used to ascertain the credit risk of individuals and, thereby, accept or reject them as borrowers. These are especially useful in the context of small-ticket loans where decisions can be taken almost instantaneously with minimal human intervention and low operational cost. The applicant's current or past data is most often compared with others having a similar profile to the applicant who have taken similar loans in the past. The applicant would be considered risky if those having similar profiles as the applicant had been found to be defaulters. However, this has its caveats since certain sections of the society might have to continuously face financial exclusion. For instance, immigrants might in general have lower repayment probability since they have (i) restrictions to accessing different facilities, (ii) lower knowledge of the local network crucial to spread their business or (iii) more stringent government regulations. If the algorithm has access to this attribute of an applicant (immigrant or not) then it might continuously reinforce the financial exclusion of the immigrants. Similar discrimination might feature based on race, ethnicity, and gender.

1.3.3 AI and the law

AI algorithms for various decision making is being increasingly used in the law 'industry'. For instance an AI system might pick up a wrong signal from the market trend thus leading to loss of capital and an eventual fraudulent bankruptcy of a company. Should the developer of the system be held guilty? A 'lawful' inspection would make it apparent that a necessary motive of the developer in this fraudulent activity is missing.

A pertinent question is whether legislators can take the liberty of defining criminal liability without a fault requirement. This would immediately mean that persons who designed or deployed the AI system regardless of whether they were aware of the potential illegal behavior are liable to be charged as responsible/guilty. There has been increasing use of such faultless liability in tort law. With all these advances and rising complexities, it is important for the law industry to evolve mechanisms to identify who is/are actually liable for the actions taken by an AI algorithm. Even if faultless liability is in use, it would be important to be precise since a 'developer' of an AI system can be a team of people rather than an individual. The responsibility might as well perpetuate to include other roles like program leaders, design experts or their superiors going all the way up to the top management.

1.3.4 Bias and fairness

Bias. AI systems are built by humans and thus the same biases as those existing in the human psyche are susceptible to perpetuate into the design of these systems. Biases can arise due to the data that is used to train the AI models or due to the biased ethical values upheld by the system developers. The number of cases showing evidence of such social bias is ever-increasing.

One of the glaring examples dates back to the study [23] by the ProPublica group which showed how black men and women were flagged with the risk of recidivism far more often by the COMPAS software than in reality. Surprisingly white men and women were flagged at less risk of recidivism than reality by the same software. Further, there have been reports about bias in the Facebook ad delivery system— highly paid job ads were shown far more frequently to men than women [38, 39]. Similarly, delivery of housing ads were biased with respect to ethnicity and race [38]. Image databases that are curated from a very small demographic and used for search purposes can produce very biased search results. For instance, such search systems frequently connect women with homemaking activities and men with white collar jobs. Queries like 'babysit' will almost always show images of women while queries like 'baseball' would only show images of men. A recent study has shown how Amazon steers the attention of the marketplace toward its own in-house products (private labels) [40]. The authors further observe that sponsored private label products draw in 50 times more recommendations than their third-party competitors.

There can also be huge inaccuracy and bias in the datasets on which AI models are trained. Many of these arise from inaccurate and biased annotations of the data points. Many FRSs, as pointed out earlier, attempt to identify the gender and age of individuals based on their facial images uploaded into the system. Predictions made by such systems are known to be highly inaccurate for certain intersectional groups (e.g., black women) [21, 41]. This disparity increases manifold if the images are noisy [42]. One particular FRS [43] that also predicts profession and personal characteristics from the image has been criticized to be offensive and racist.

Many of these problems are exacerbated due to the 'black-box' nature of the algorithms that work at the back-end. This makes it impossible for the third-party to assess how the predictions are being made and what special features used in these algorithms result in biased predictions. Thus there is a need to make AI algorithms fair, transparent, and accountable which also resonates with the central objective of this book.

Fairness. As more and more decision-making tasks are being delegated to AI systems such as shortlisting of CVs, predicting the chances of recidivism, credit scoring, or university admissions, there is an urgent need to make these systems fair and impartial. Care needs to be taken so that the decisions made by these systems do not intensify the already existing and very harmful social divide.

The important research question, therefore, is how to make algorithms fair. There seems to be no straightforward answer to this question [44]. The main reason is that it is impossible to understand what clues the complex neural network based models pick up from the training data to make its predictions. For instance, in the case of COMPAS, even if the algorithm is never fed with the race information, it makes an inference of the same based on some other seemingly benign attribute such as location/zip code.

One way to operationalize fairness could be to ask what a fair outcome should look like. In this context, the notions of discrimination are important. The first type of discrimination is called *disparate treatment* [45] in which similarly situated individuals are differentially treated. For instance, if an employer treats some individuals more favorably than other similarly situated individuals based on religion, race, caste, gender, etc, then it constitutes a case of disparate treatment. When the discrimination is unintentional it is called *disparate impact* [45]. Such cases arise due to policies and practices that are apparently neutral but eventually result in disproportionate impact on a particular group of individuals. For instance if job applicants for a particular company are recruited based on their weight carrying capacity while there is no requirement of such a quality for the job advertised then this will be discriminative toward people having low weight carrying capacity. A third type of discrimination arised when the misclassification rates across the different classes of individuals (e.g., white versus black) are different and is known as *disparate mistreatment* [46]. For instance, in the COMPAS example discussed earlier, there is a disparate mistreatment since the misclassification rates of recidivism for African-American individuals are found to be more than white Americans.

The objective is to devise/alter machine learning algorithms in such a way that the above types of discrimination are minimized. The most common technique is to change or constrain the learning objective/loss function to forbid the emergence of such discrimination in the predictions of the algorithm. However, this has its own challenges. First, having such a change or constraint might make the function lose some of its desirable properties (e.g., convexity). Further, the notions of discrimination and therefore fairness might conflict with one another and make it impossible to design functions that eliminate all types of discrimination. We shall devote the next chapter to the discussion of all these nuanced issues.

1.4 What is a computational perspective?

The primary departure that this book attempts to make is to bring on board a computational perspective. While many books have been published which are thematically similar, what is lacking in them is this perspective. By computational perspective, we mean a computer scientists' view on the issue of AI and ethics. In particular, this book will have its roots in the algorithmic design principles that underpin the developments of this field. In the fair machine learning subpart, the book will discuss aspects of redesigning the optimization functions for classification and regression problems to tackle disparate treatment, disparate impact, and disparate mistreatment. This part will also present algorithms to debias word vector representations that are routinely used in natural language processing applications. In the content governance chapter (chapter 4), this book will extensively cover concepts of algorithmic auditing of various software systems (e.g., FRSs) in both normal and adversarial settings. The next part of this same chapter will deal with algorithms to detect and mitigate hate speech, online violence, and misinformation. In the final part of this chapter we will focus on algorithms for the detection of political inclinations of individuals and biases across media houses. In the next chapter (chapter 5), the book will describe the algorithms to understand the inner workings of complex deep neural models. We shall cover two types of concepts—interpretability and explainability. In the interpretability part we shall discuss what features are most important in a model prediction. In the explainability part, assuming the complex model to be a black-box we will discuss algorithmic techniques that can be used to build simple proxy white-box models to explain the predictions of the black-box models. The last chapter of this book (chapter 6) will talk about the algorithmic challenges in developing autonomous vehicles and the ways to encode moral values into an AI algorithm. We shall also point out how AI algorithms interact with the law industry. For instance, if AI algorithms are tasked to make law decisions, how should the law industry adapt itself; should new laws be passed and many old laws be abolished? Another issue that we shall discuss is the rising issue of data protection and privacy which are closely related to ethics. We shall end with a note on the impact of singularity and superintelligence.

References

[1] Russell S J and Norvig P 2009 *Artificial Intelligence: A Modern Approach* 4th edn (Upper Saddle River, NJ: Prentice Hall)

[2] MacQueen J 1967 Classification and analysis of multivariate observations *5th Berkeley Symposium on Mathematical Statistics and Probability* (Berkeley, CA: University of California Press) pp 281–97

[3] Kaufman L and Rousseeuw P J 2009 *Finding Groups in Data: An Introduction to Cluster Analysis* (New York: Wiley)

[4] Pelleg D and Moore A W 2000 X-means: extending K-means with efficient estimation of the number of clusters *ICML'00: Proc. 17th Int. Conf. on Machine Learning* vol 1 pp 727–34

[5] Ester M, Kriegel H-P, Sander J and Xu X 1996 A density-based algorithm for discovering clusters in large spatial databases with noise *Proc. 2nd Int. Conf. on Knowledge Discovery and Data Mining, KDD'96* (Washington, DC: AAAI Press) pp 226–31

[6] Pearson K 1901 LIII. On lines and planes of closest fit to systems of points in space *Lond. Edinb. Dublin Philos. Mag. J. Sci.* **2** 559–72

[7] Wold S, Esbensen K and Geladi P 1987 Principal component analysis *Chemom. Intell. Lab. Syst* **2** 37–52

[8] Stewart G W 1993 On the early history of the singular value decomposition *SIAM Rev.* **35** 551–66

[9] Harman H H 1976 *Modern Factor Analysis* (Chicago, IL: University of Chicago Press)

[10] Bingham E and Mannila H 2001 Random projection in dimensionality reduction: applications to image and text data *Proc. 7th ACM SIGKDD Int. Conf. on Knowledge Discovery and Data Mining* pp 245–50

[11] Vempala S S 2005 *The Random Projection Method* vol 65 (Providence, RI: American Mathematical Society)

[12] Krizhevsky A, Sutskever I and Hinton G E 2012 Imagenet classification with deep convolutional neural networks *Advances in Neural Information Processing* vol 25 ed F Pereira, C J Burges, L Bottou and K Q Weinberger (Red Hook, NY: Curran Associates, Inc.) https://proceedings.neurips.cc/paper/2012/file/c399862d3b9d6b76c8436e924a68c45b-Paper.pdf

[13] Szegedy C, Liu W, Jia Y, Sermanet P, Reed S, Anguelov D, Erhan D, Vanhoucke V and Rabinovich A 2015 Going deeper with convolutions *Proc. IEEE Conf. on Computer Vision and Pattern Recognition* pp 1–9

[14] Lin M, Chen Q and Yan S 2013 *Network in Network* Network in network *2nd Int. Conf. on Learning Representations, {ICLR} 2014 (Banff, AB, Canada, April 14–16, 2014)*

[15] He K, Zhang X, Ren S and Sun J 2016 Deep residual learning for image recognition *Proc. IEEE Conf. on Computer Vision and Pattern Recognition* pp 770–8

[16] Hochreiter S and Schmidhuber J 1997 Long short-term memory *Neural Comput.* **9** 1735–80

[17] Cho K, Van Merriënboer B, Bahdanau D and Bengio Y 2014 *On the Properties of Neural Machine Translation: Encoder–Decoder Approaches* (arXiv:1409.1259 [cs.CL])

[18] Devlin J, Chang M-W, Lee K and Toutanova K 2019 BERT: pre-training of deep bidirectional transformers for language nderstanding *Proc. 2019 Conf. of the North American Chapter of the Association for Computational Linguistics: Human Language Technologies, vol 1 (Long and Short Papers)* vol 1 (Minneapolis, MN: Association for Computational Linguistics) pp 4171–86

[19] Saumya R 2021 *Alexa: A Catalyst in the Evidence Law?* https://criminallawstudiesnluj.wordpress.com/2021/09/18/alexa-a-catalyst-in-the-evidence-law/

[20] Paul D 2017 *Your Own Pacemaker Can Now Testify Against You in Court* https://www.wired.com/story/your-own-pacemaker-can-now-testify-against-you-in-court/

[21] Buolamwini J and Gebru T 2018 Gender shades: intersectional accuracy disparities in commercial gender classification *Conf. on Fairness, Accountability and Transparency* pp 77–91 https://proceedings.mlr.press/v81/buolamwini18a.html

[22] Andrejevic M and Selwyn N 2020 Facial recognition technology in schools: critical questions and concerns *Learn. Media Technol.* **45** 115–28

[23] Angwin J, Larson J, Mattu S and Kirchner L 2016 *Machine bias Ethics of Data and Analytics* https://www.propublica.org/article/machine-bias-risk-assessments-in-criminal-sentencing

[24] Gillespie T 2018 *Custodians of the Internet: Platforms, Content Moderation, and the Hidden Decisions that Shape Social Media* (New Haven, CT: Yale University Press)

[25] Halevy A, Canton-Ferrer C, Ma H, Ozertem U, Pantel P, Saeidi M, Silvestri F and Stoyanov V 2022 Preserving integrity in online social networks *Commun. ACM* **65** 92–8

[26] Ryan D *et al* 2020 *AI Advances to Better Detect Hate Speech* https://ai.facebook.com/blog/ai-advances-to-better-detect-hate-speech/

[27] Guo C, Cao J, Zhang X, Shu K and Yu M 2019 Exploiting Emotions for Fake News Detection on Social Media (arXiv:1903.01728v4 [cs.CL])

[28] Güera D and Delp E J 2018 Deepfake video detection using recurrent neural networks *2018 15th IEEE Int. Conf. on Advanced Video and Signal Based Surveillance (AVSS)* (Piscataway, NJ: IEEE) pp 1–6

[29] Wei Y, Singh L and Martin S 2016 Identification of extremism on Twitter *2016 IEEE/ACM Int. Conf. on Advances in Social Networks Analysis and Mining (ASONAM)* (Piscataway, NJ: IEEE) pp 1251–5

[30] Gaikwad M, Ahirrao S, Phansalkar S and Kotecha K 2021 Online extremism detection: a systematic literature review with emphasis on datasets, classification techniques, validation methods, and tools *IEEE Access* **9** 48364–404
[31] Wilson T, Wiebe J and Hoffmann P 2005 Recognizing contextual polarity in phrase-level sentiment analysis *Proc. Human Language Technology Conf. and Conf. on Empirical Methods in Natural Language Processing* pp 347–54 https://aclanthology.org/H05-1044
[32] Kosinski M 2021 Facial recognition technology can expose political orientation from naturalistic facial images *Sci. Rep.* **11** 1–7
[33] Harding L 2018 *Twitter Admits far More Russian Bots Posted on Election than It Had Disclosed* https://www.theguardian.com/technology/2018/jan/19/twitter-admits-far-more-russian-bots-posted-on-election-than-it-had-disclosed
[34] Kirkpatrick D D 2017 *Signs of Russian Meddling in Brexit Referendum* https://www.nytimes.com/2017/11/15/world/europe/russia-brexit-twitter-facebook.html
[35] Mohan M 2017 *Macron Leaks: The Anatomy of a Hack* https://www.bbc.com/news/blogs-trending-39845105
[36] Confessore N 2018 *Cambridge Analytica and Facebook: The Scandal and the Fallout so Far* https://www.nytimes.com/2018/04/04/us/politics/cambridge-analytica-scandal-fallout.html
[37] Campisi N and Lupini C 2021 From Inherent Racial Bias to Incorrect Data–The Problems with Current Credit Scoring Models https://www.forbes.com/advisor/credit-cards/from-inherent-racial-bias-to-incorrect-data-the-problems-with-current-credit-scoring-models/
[38] Ali M, Sapiezynski P, Bogen M, Korolova A, Mislove A and Rieke A 2019 Discrimination through optimization: how Facebook's ad delivery can lead to biased outcomes *Proc. ACM on Human–Computer Interaction* vol 3 CSCW, pp 1–30
[39] Kaplan L, Gerzon N, Mislove A and Sapiezynski P 2022 Measurement and analysis of implied identity in ad delivery optimization *Proc. 22nd ACM Internet Measurement Conf.* pp 195–209
[40] Dash A, Chakraborty A, Ghosh S, Mukherjee A and Gummadi K P 2021 When the umpire is also a player: bias in private label product recommendations on e-commerce marketplaces *Proc. 2021 ACM Conf. on Fairness, Accountability, and Transparency* pp 873–84
[41] Raji I D, Gebru T, Mitchell M, Buolamwini J, Lee J and Denton E 2020 Saving face: investigating the ethical concerns of facial recognition auditing *Proc. AAAI/ACM Conf. on AI, Ethics, and Society* pp 145–51
[42] Jaiswal S, Duggirala K, Dash A and Mukherjee A 2022 Two-face: adversarial audit of commercial face recognition systems *Proc. Int. AAAI Conf. on Web and Social Media* vol 16 pp 381–92
[43] Chu W-T and Chiu C-H 2016 Predicting occupation from images by combining face and body context information *ACM Trans. Multimedia Comput. Commun. Appl.* **13** 1–21
[44] Barocas S, Hardt M and Narayanan A 2019 *Fairness and Machine Learning: Limitations and Opportunities* fairmlbook.org http://www.fairmlbook.org
[45] Barocas S and Selbst A D 2016 Big data's disparate impact *Calif. Law Rev.* **104** 671–732
[46] Zafar M B, Valera I, Rodriguez M G and Gummadi K P 2017 Fairness beyond disparate treatment and disparate impact: learning classification without disparate mistreatment *Proc. 26th Int. Conf. on World Wide Web* pp 1171–80

IOP Publishing

AI and Ethics
A computational perspective
Animesh Mukherjee

Chapter 2

Discrimination, bias, and fairness

2.1 Chapter foreword

This chapter will first introduce the concepts necessary for designing fair machine learning algorithms. This will be followed by a detailed account of how such concepts are formalized, assimilated, and operationalized into state-of-the-art machine learning algorithms to make them fair. In doing so, we shall start by defining what are *sensitive attributes*. This is one of the key points to understand because most of the bias, discrimination, and societal exclusion revolve around this point. Once we have a working definition of the sensitive attributes, we shall introduce the widely known types of discrimination against a protected class of individuals who bear one or more of the sensitive attributes. As a follow-up, we shall put in context some critical types of biases that feature in different real-life situations. While some of these bias types are very apparent, others are quite nuanced and need deeper levels of investigation to ascertain their presence. Next, we quantify the notions of fairness that attempt to ameliorate discrimination/bias in various systems. Finally, we detail the different ways in which fairness notions are incorporated into the pipeline of traditional machine learning algorithms. We identify the mathematical difficulties and ways to circumvent them in standard classification, regression, and various other learning setups.

2.2 Sensitive attributes

In the different learning setups that we discussed in the previous chapter, differentiating or categorizing an instance or a group of instances is the primary objective in most cases. With the recent prominence of these algorithms, the instances classified or differentiated are often human beings. Sometimes the aforementioned differentiation may be unjustified and done on the basis of attributes that do not have any practical or moral relevance. For example race, gender, and sexual orientation of an individual may be considered practically irrelevant for a machine learning algorithm predicting whether to hire someone or not. Even though such features may have some underlying statistical relevance due to the historical stereotypes prevalent in society, using such features may not be morally acceptable. Acknowledging the sensitivity of such attributes, in the literature, these are more commonly referred to as *sensitive attributes*. Sensitive attributes may be earmarked based on legal regulations or doctrines of a country or due to the organizational values of a company. Some of the common examples of sensitive attributes include race, ethnicity, gender, caste, age, disability, religion, marital, and financial status.

The aforementioned list of sensitive features is not completely arbitrary either. Rather, it stems from existing regulations which forbid differentiating individuals and groups based on these socially salient attributes that have served as a basis for unjustified and systematic adverse treatment through the course of our history. To this end, an individual associated with one or more of these attributes (socially salient groups who have been historically discriminated against) is said to belong to the protected class. Typically, a protected class stems from the categorization of data into groups that may result in direct discrimination due to the use of such group information. For instance, criminal ethnicity should not be used to determine the risk of recidivism. In this case, the numerical encoding of ethnic groups like African-Americans or Caucasians in the context of the Western world may lead to bias and discrimination against the historically oppressed group (i.e., the protected group/class). Some of the well-known conventions and laws that guide the definition of a sensitive attribute are listed below.

- Discrimination in housing. The US Fair Housing Act prohibits any form of discrimination based on race, ethnicity, gender, etc, in housing opportunities.
- In the US, it is prohibited to make hiring decisions based on different sensitive attributes like race and gender. A cluster of laws also known as the Federal Equal Employment Opportunity—Civil Rights Act Title VII 1964, EPA 1963, ADEA 1967, ADA 1990, Rehabilitation Act 1973, Civil Rights Act 1991, GINA 2008—ensures that companies adhere to this guideline.
- One of the recent additions to the above guidelines is ECOA which directs firms for testing automatic recruitment/shortlisting algorithms to be fair. Unfair outcomes from such algorithms are subject to penalties.
- The UN General Assembly resolutions 1965 prohibit all forms of racial discrimination. Similarly, the UN General Assembly resolutions 1979 prohibit any form of discrimination against women. In the 2006 UN General Assembly, a resolution was passed to preserve the rights of persons with disabilities.
- Many countries have coded laws around sensitive attributes in their constitution. For instance, Article 17 of the Indian constitution bans the practice of 'untouchability' of any form. Further the Scheduled Castes and the Scheduled Tribes (Prevention of Atrocities) Act, 1989 prevents any sort of caste-based discrimination in hiring, educational opportunity, access to public properties, etc, against the low caste population in India.

2.3 Notions of discrimination

While it is very hard to have a single definition of discrimination, there has been a lot of research to formally represent the concept. At least three of the following types of notions of discrimination have become widely popular. Note that in all these cases, the discrimination is against a protected class bearing one or more of the sensitive attributes (as defined in the previous section).

2.3.1 Disparate treatment

Disparate treatment is a form of discrimination where similarly situated individuals are differently treated. There are two types of disparate treatment—*formal* and *intentional* [1].

Formal discrimination refers to the direct denial of opportunities based on protected class membership and a curious reference to the concept of *rational racism*. The main thesis here is that one should believe what is rational and therefore whatever is rational should also be rightful. However, one has to understand that although there are many existing policies that look superficially reasonable, they are actually morally wrong. This connects back to the concept of 'naturalistic fallacy' which argues that if something is natural it must be good/morally correct. For instance, such a fallacious argument as the following—during a disaster all old and disabled people should be left to die as the cost to rescue them is far larger than the foreseeable future benefit they would bring to society—would be considered rightful. Almost every one of us would find the above advocacy to be appalling. Similarly, many things that look rational possibly based on empirical evidence may not justify a particular action to be good. One of the glaring examples is the observation that African-Americans are genetically or culturally prone to be criminals merely because of their overrepresentation in jails.

Intentional discrimination is best understood through the 'burden-shifting' scheme of events. Within this scheme, the victim has the initial responsibility to establish a case for discrimination against him or her. This is usually done by gathering evidence that a similarly situated individual not belonging to the protected class would not have suffered a similar discrimination. One of the glaring examples appears in the form of irregularities in job hiring and promotions. If the plaintiff is able to establish the case of discrimination then the burden shifts to the defendant to procure a satisfactory account of non-discriminatory basis of the visible irregularity. Note that the defendant does not have the liability to prove that the basis is true. The burden now comes back to the plaintiff to prove that the account provided by the defendant is only a form of excuse and is therefore not legitimate.

2.3.2 Disparate impact

In scenarios where discrimination manifests as an outcome even when there is no discriminatory intent, it becomes a case of disparate impact. Once again considering the example of job hiring, the defendant can make a case that the apparently observed discriminatory outcome manifests from a job-related requirement or a specific necessity for running the business. In such a scenario the plaintiff can still win the case by providing evidence of an alternative practice that is potentially more non-discriminatory.

2.3.3 Disparate mistreatment

There is one caveat with the disparate impact doctrine; when the historical ground-truth decisions are available, discriminatory outcomes might be justified based on the ground-truth. In such scenarios, the alternative notion of disparate mistreatment is more useful. A decision-making process is said to suffer from disparate mistreatment if the misclassification rates are different for the protected class as compared to the other classes. Different kinds of misclassifications become important based on the application at hand. For instance, in some applications the false positive rate is more important while in others the false negative rate takes precedence. Consider the case of assessing the risk of recidivism; in this particular scenario it is important to balance the false positive rate across the classes (e.g., classes based on different races) since it is better to let a guilty person go than accuse an innocent one. On the other hand, in the context of credit scoring it is important to equalize the false negative rates so that deserving candidates from a protected class are not discriminated.

2.4 Types of bias

Bias is a strong and preconceived notion about someone or something that is built from our knowledge/perception or the lack thereof of that individual or thing. Understanding biases is a key component of scientific literacy. This is because notwithstanding the educational background, parental upbringing, or intellectual ability, one can always become susceptible to biases. Some common forms of biases are noted below.

2.4.1 Confirmation bias

Confirmation bias is the tendency to only believe in or search for information that supports one's preconceived beliefs. Thus, the individual avoids believing any contrary views or opinions. An example of this is often observed in support of politicians during elections, where the individual only accepts positive opinions and refuses to accept or believe in anything that paints their favored candidate in a bad light.

2.4.2 Population bias

Population bias occurs when certain characteristics like demographics or representative groups in the user population differ from the original target population. For example, face detection software in cameras of the early 2010s had problem in detecting non-Caucasian faces [2]. When people tried to take photographs of people of Asian demographics, Nikon Coolpix S630 digital camera used to show an error message that reads *'Did someone blink?'*.

2.4.3 Sampling/selection bias

Selection or sampling bias occurs when the target population is sampled in a non-random manner. As some groups are represented more frequently, trend prediction may not be correct or generalizable for the entire population. An example of this bias arises when the prediction trends get reversed if we attempt to use a model trained on one dataset (where a subgroup might be under-represented) for another similar dataset (where such under-representation does not exist for the given subgroup).

2.4.4 Interaction bias

Interaction bias results from the way a user interacts with a system. For example, due to our cognitive biases most of us read from left to right and from top to bottom. Hence when search results are presented to users usually the results toward the top-left corner get disproportionately higher attention [3]. Thus such biases are often a compounding effect of the way the results or choices are presented and users' cognitive biases while interacting with them.

2.4.5 Decline bias

Decline bias is the outcome of comparison between the past and present and concluding that the present is always getting increasingly worse. This may be due to the inability of the individual to deal with changes taking place in the present.

2.4.6 The Dunning–Kruger effect

The Dunning–Kruger effect is a well-known form of cognitive bias wherein a person with limited knowledge of or skill in a subject considers themselves to be an expert on the same and overestimates their abilities to perform the task successfully.

2.4.7 Simpson's paradox

Simpson's paradox refers to the situation where trends observed in the subgroups of a dataset differ from those observed at the aggregate level. One of the interesting examples of this bias was when a lawsuit was filed against the UC Berkeley graduate admissions reporting gender based discrimination [4]. It was alleged that women were admitted in smaller fractions than men for graduate studies. However, when the data was analysed department wise it was observed that women applicants were in fact at a slightly more advantageous position. The misunderstanding resulted from the fact that women candidates typically applied more to departments where the admission rate was lower for both women and men.

2.4.8 Hindsight bias

This type of bias arise in match/lottery outcome predictions. Such biases make one prone to overestimate their predictive/forecasting ability [5]. People often tend to construct a plausible explanation for the consequence that had occurred, and start believing that they can correctly predict the outcomes of other events beforehand. Often the information on which they would base these predictions would lead them to make a wrong forecast.

2.4.9 In-group bias

In-group bias [6] is the result of belief and/or support for members from one's own social group instead of an outsider. For example, an employer may attempt to recruit people from their own social group without objectively evaluating their competence for the role.

2.4.10 Emergent bias

Emergent bias 'emerges' in a working system while real users actually start interacting regularly with the system [7]. This bias maybe borne from a change in the population, cultural differences, or awareness of new/additional societal knowledge usually after the system has been launched for some time. These types of biases are routinely observed in various HCI designs.

2.5 Operationalizing fairness

In order to make machine learning models fair, the concept of fairness need to be operationalized. In other words, we need to have mathematical definitions of fairness that have to be developed from the existing notions of fairness. Some of the most popular definitions are listed below. In all these definitions, we assume a binary classification setting (which can be easily generalized to a multi-class setting). We shall denote the sensitive attribute by z, i.e., if $z = 1$, we assume the individual to be in the protected class. Further let $\hat{y}$ be the predicted label and y be the ground-truth label for the classification task.

2.5.1 Demographic/statistical parity

A predictor is said to maintain statistical or demographic parity if $p(\hat{y}|z = 0) = p(\hat{y}|z = 1)$. This means that the likelihood of a positive outcome should be the same irrespective of whether an individual belongs to the protected class or not. In terms of the confusion matrix the positive rate (p_+, which includes both the true and the false positive rates) needs to be the same for the two classes, i.e., $p_+(z = 0) = p_+(z = 1)$. In reality, it might be not possible to have the positive rates exactly equal for both the classes, however, the objective should be to bring them as close as possible.

However, researchers often point out some practical irrelevance in this basic definition of fairness. First, it enables a system to be lazy wherein if the system predicts positive outcomes correctly for one of the groups and randomly selects equal proportion of candidates from the other group—it still ensures demographic parity. Nevertheless, it may potentially be detrimental for the downstream application.

2.5.2 Equal opportunity

Equal opportunity on the other hand was formulated to circumvent the aforementioned loophole in statistical parity. It mandates that $p(\hat{y} = 1|z = 0, y = 1) = p(\hat{y} = 1|z = 1, y = 1)$. This means that the two classes $z = 0$ and $z = 1$ should have the same true positive rates, i.e., a person belonging to the positive class should be correctly predicted to be in the positive class irrespective of the value of z. Once again, exact equality is not the objective; rather, the values should be as close to one another as possible.

2.5.3 Predictive parity

Predictive parity mandates that the positive predictive rates (PPV) for both $z = 0$ and $z = 1$ should be the same. PPV stands for the probability that the individuals that are predicted to be in the positive class are actually in the positive class. In other words, $p(y = 1|z = 0, \hat{y} = 1) = p(y = 1|z = 1, \hat{y} = 1)$.

2.5.4 Equalized odds

Equalized odds mandates that $p(\hat{y} = 1|z = 0, y \in 0, 1) = p(\hat{y} = 1|z = 1, y \in 0, 1)$. In other words, both the positive and the negative rates should be the same (for practical purposes, as close as possible) irrespective of the value of z.

2.5.5 Disparate mistreatment

A predictor is said to avoid disparate mistreatment if the misclassification rates are the same irrespective of the value of z. It mandates that $p(\hat{y} \neq y|z = 0) = p(\hat{y} \neq y|z = 1)$. In most applications, the objective is to bring both the false positive and the false negative rates close for both the classes.

2.5.6 Fairness through (un)awareness

Fairness through awareness. The key concept here is that similar individuals should have similar predicted outcomes. This means that if two individuals perform equally on a predefined similarity metric, the predictor should predict the same outcome for both of them.

Fairness through unawareness. The main idea here is that no protected attribute should be used by the predictor to decide the outcome.

2.6 Relationships between fairness metrics

In this section we question whether these metrics can be simultaneously satisfied by a predictor. In order to do this we would need to resort to a few definitions as follows.

Base rate. For a particular group the base rate (b_z) is defined as the number of members in the group belonging to the positive class expressed as a fraction of the total number of members in the group. Therefore for $z = 0$ and $z = 1$ if the base rates are different then we have $p(y = 1|z = 0) \neq p(y = 1|z = 1)$.

True positive rate. The true positive rate (TPR) for a group z can be defined as $\text{TPR}_z = p(\hat{y} = 1|y = 1, z \in 0, 1)$.

False positive rate. The false positive rate (FPR) for a group z can be defined as $\text{FPR}_z = p(\hat{y} = 1|y = 0, z \in 0, 1)$.

Positive predictive rate. The positive predictive rate (PPV) for a group z can be defined as $\text{PPV}_z = p(y = 1|\hat{y} = 1, z \in 0, 1)$.

2.6.1 Connection between statistical parity, predictive parity, and equalized odds

The basic probability relation allows us to draw connections between the three concepts—statistical parity, predictive parity, and equalized odds. To situate this, from the law of conditional probability we know that $p(x, y) = p(x|y)p(y) = p(y|x)p(x)$. Thus we have

$$p(y, \hat{y}|z) = p(y|\hat{y}, z)p(\hat{y}|z) = p(\hat{y}|y, z)p(y|z) \qquad (2.1)$$

Note that here $p(y|\hat{y}, z)$ is the predictive parity, $p(\hat{y}|z)$ is the statistical parity, $p(\hat{y}|y, z)$ is the equalized odds and $p(y|z)$ is the base rate. In the next three sections we discuss the pairwise interrelations between these three metrics.

2.6.2 Statistical and predictive parity

We analyze the case where statistical parity and predictive parity can both hold for a predictor. Recall that statistical parity ensures $p(\hat{y} = 1|z = 0) = p(\hat{y} = 1|z = 1)$ or $p(\hat{y} = 1)$. The difference in the PPV for the two groups should be

$$p(y = 1|z = 0, \hat{y} = 1) - p(y = 1|z = 1, \hat{y} = 1)$$
$$= \frac{\text{TPR}_0 p(y = 1|z = 0) - \text{TPR}_1 p(y = 1|z = 1)}{p(\hat{y} = 1)} \qquad (2.2)$$

If predictive parity has to hold then the LHS has to be zero. This means that

$$\frac{\text{TPR}_0}{\text{TPR}_1} = \frac{p(y = 1|z = 1)}{p(y = 1|z = 0)} = \frac{b_1}{b_0} \qquad (2.3)$$

This means that predictive parity would hold when the ratio of the true positive rates of the two groups is equal to the inverse of the ratio of the base rates of the two groups. This means that statistical parity and predictive parity can hold together even when the base rates for the two groups are different; however, if the ratio of the base rates is very far from 1 then the true positive rates of one of the classes need to be much worse than the other and therefore such a predictor is of very limited utility.

2.6.3 Statistical parity and equalized odds

In this section we discuss whether statistical parity and equalized odds could both be simultaneously satisfied. In this case, we would need $\text{TPR} = \text{TPR}_0 = \text{TPR}_1$ and $\text{FPR} = \text{FPR}_0 = \text{FPR}_1$ for equalized odds. Now we can write

$$p(\hat{y} = 1|z) = \sum_y p(\hat{y} = 1|y, z)p(y|z)$$
$$= p(\hat{y} = 1|y = 1, z)p(y = 1|z)$$
$$+ p(\hat{y} = 1|y = 0, z)p(y = 0|z)$$
$$= \text{TPR}\, p(y = 1|z) + \text{FPR}\, p(y = 0|z),$$
$$\because \quad \text{TPR} = \text{TPR}_{0/1},\ \text{FPR} = \text{FPR}_{0/1}$$

$$\Rightarrow p(\hat{y} = 1|z = 0) - p(\hat{y} = 1|z = 1)$$
$$= (\text{TPR} - \text{FPR})(p(y = 1|z = 0) - p(y = 1|z = 1)) \quad (2.4)$$

For statistical parity to hold, the LHS has to be equal to zero. This means that either TPR − FPR = 0 or $p(y = 1|z = 0) - p(y = 1|z = 1) = 0$. The latter case could be ruled out since it disallows the base rates for the two groups to be different (which is often the case). The only viable option, therefore, is that TPR = FPR for the statistical parity to hold. Thus, although statistical parity and equalized odds can hold simultaneously the predictor has to have TPR = FPR which makes it practically unusable.

2.6.4 Predictive parity and equalized odds

When both equalized odds and predictive parity need to hold, one has to ensure $\text{TPR}_0 = \text{TPR}_1$, $\text{FPR}_0 = \text{FPR}_1$, and $\text{PPV}_0 = \text{PPV}_1$. As obtained in the previous section we have

$$p(\hat{y} = 1|z) = \text{TPR}_0 p(y = 1|z) + \text{FPR}_0 p(y = 0|z) \quad (2.5)$$

Now combining equations (2.1) and (2.5) we can write

$$p(\hat{y} = 1|y = 1, z = 0)p(y = 1|z = 0)$$
$$= p(y = 1|\hat{y} = 1, z = 0)(\text{TPR}_0 p(y = 1|z = 0) + \text{FPR}_0 p(y = 0|z = 0))$$

$$\Rightarrow \text{TPR}_0 p(y = 1|z = 0) = \text{PPV}_0(\text{TPR}_0 p(y = 1|z = 0) + \text{FPR}_0 p(y = 0|z = 0))$$

$$\Rightarrow \text{TPR}_0 p(y = 1|z = 0) = \text{PPV}_0(\text{TPR}_0 p(y = 1|z = 0) + \text{FPR}_0(1 - p(y = 1|z = 0)))$$

$$\Rightarrow p(y = 1|z = 0) = \frac{\text{PPV}_0 \text{FPR}_0}{\text{PPV}_0 \text{FPR}_0 + (1 - \text{PPV}_0)\text{TPR}_0} \quad (2.6)$$

In a similar fashion we can obtain

$$\Rightarrow p(y = 1|z = 1) = \frac{\text{PPV}_1 \text{FPR}_1}{\text{PPV}_1 \text{FPR}_1 + (1 - \text{PPV}_1)\text{TPR}_1} \quad (2.7)$$

Now since we started with the assumption that $\text{TPR}_0 = \text{TPR}_1$, $\text{FPR}_0 = \text{FPR}_1$, and $\text{PPV}_0 = \text{PPV}_1$, so equations (2.6) and (2.7) should be exactly the same. This means that for equalized odds and predictive parity to hold together, the base rates have to be identical. This, for all practical purposes, is never the case.

Note that in the case of equal opportunity (a lenient version of equalized odds), the condition $\text{FPR}_0 = \text{FPR}_1$ can be removed from equations (2.6) and (2.7) which allows for unequal base rates and can simultaneously hold with predictive parity.

2.7 Individual versus group fairness

The definitions of fairness outlined above are applicable on different types population groupings. These are as follows.

Individual fairness. Individual fairness mandates ensuring similar predictions for similar individuals. The fairness through awareness and the fairness through unawareness measures are the most suitable measures for ensuring individual fairness.

Group fairness. Group fairness mandates treating different groups equally. Demographic parity, predictive parity, equal opportunity, equalized odds, and avoiding disparate mistreatment are all applicable here.

Subgroup fairness. Subgroup fairness attempts to achieve the best of both individual and group fairness criteria. It chooses a group fairness constraint and then requires that this constraint is ensured across an infinitely many number subgroups.

Practice problems

Q1. Write the mathematical expressions for the probability function of *equalized odds* and *equal opportunity* with explanations for all the variables used. Briefly explain the relationship between the two.

Q2 (a). *Disparate mistreatment* is defined as follows—'A decision-making process is said to be suffering from disparate mistreatment with respect to a given sensitive attribute (e.g., race) if the misclassification rates differ for groups of people having different values of that sensitive attribute (e.g., blacks and whites)'. Gotham City's police department has decided to automate the process of identifying *persons of interest* who should be stopped and searched based on the suspicion of possessing illegal weapons. They have decided to use three classifiers, C_1, C_2, and C_3 for this job. Each of the classifiers makes its decision based on three features—*gender*, *clothing bulge*, and *proximity to the crime scene*. Out of these three, *gender* is a sensitive feature. Figure 2.1 has the values for each attribute as well as the ground-truth and each classifier's decision.
Answer the following—
1. Calculate the FPR (false positive rate) for each classifier for both the male and female gender.
2. Calculate the FNR (false negative rate) for each classifier for both the male and female gender.
3. Which of the classifiers, if any, are unfair due to disparate mistreatment? Explain.

Q2 (b). Answer the following based on figure 2.1.
1. Which of the classifiers, if any, are unfair due to disparate treatment? Explain.
2. Which of the classifiers, if any, are unfair due to disparate impact? Explain.

Q3. *Casual conversations* [8] is a dataset proposed by researchers from Meta Research for benchmarking AI models. The dataset is composed of 3011 subjects and contains over 45 000 videos, with an average of 15 videos per person. The videos were recorded in multiple US states with a diverse set of adults of various ages,

User Attributes			Ground Truth (Has Weapon)	Classifier's Decision to Stop		
Sensitive	Non-Sensitive					
Gender	Clothing Bulge	Prox. Crime		C_1	C_2	C_3
Female 1	1	1	No	1	1	1
Female 2	0	0	Yes	1	1	0
Female 3	1	0	Yes	1	0	1
Male 1	1	1	Yes	1	1	0
Male 2	1	0	No	1	0	1
Male 3	0	1	Yes	0	1	1

Figure 2.1. Sensitive and non-sensitive user attributes, ground-truth, and the decision made by the three classifiers.

Table 2.1. Ground truth label distribution for Facebook's casual conversations dataset [8]. T: Type.

Age	N/A (2.1%) 46–85 (29.8%) 31–45 (32.5%) 18–30 (35.6%)
Gender	Male (43.4%) Female (54.5%) Other (2.1%)
Skin tone	T I (4.0%) T II (28.3%) T III (22.9%) T IV (8.4%) T V (15.8%) T VI (20.7%)
Lighting	Dark (14.9%) Bright (85.1%)

gender, and apparent skin tone groups. The skin tone is recorded as per the Fitzpatrick scale. The distribution of the ground-truth labels is shown in table 2.1. Is the dataset, as shown in table 2.1, *fair*? Why/why not? If you think it is unfair, also answer how you can make it fair for use.

References

[1] Barocas S and Selbst A D 2016 Big data's disparate impact *Calif. Law Rev.* **104** 671–732
[2] Rose A 2010 Are face-detection cameras racist? http://content.time.com/time/business/article/0,8599,1954643,00.html
[3] Baeza-Yates R 2018 Bias on the web *Commun. ACM* **61** 54–61
[4] Bickel P J, Hammel E A and O'Connell J W 1975 Sex bias in graduate admissions: data from Berkeley: measuring bias is harder than is usually assumed, and the evidence is sometimes contrary to expectation *Science* **187** 398–404
[5] Roese N J and Vohs K D 2012 Hindsight bias *Perspect. Psychol. Sci.* **7** 411–26
[6] Taylor D M and Doria J R 1981 Self-serving and group-serving bias in attribution *J. Soc. Psychol.* **113** 201–11

[7] Friedman B and Nissenbaum H 1996 Bias in computer systems *ACM Trans. Inf. Syst.* **14** 330–47
[8] Hazirbas C, Bitton J, Dolhansky B, Pan J, Gordo A and Ferrer C C 2021 Casual conversations: a dataset for measuring fairness in AI *Proc. IEEE/CVF Conf. on Computer Vision and Pattern Recognition* pp 2289–93

Chapter 3

Algorithmic decision making

3.1 Chapter foreword

This chapter motivates the need for fair decision-making algorithms and how to design them. We begin by discussing two important case studies—the first case study describes the use of specialized software in the criminal justice system to predict the risk of recidivism based on the past records of a defendant; the second case study describes the discrimination in facial recognition systems against certain gender and/or racial groups. These are followed by a discussion on how to modify existing machine learning models to include fairness constraints. The last part of the chapter discusses how to incorporate fairness objectives in the generation of word embeddings that are routinely used in various natural language processing (NLP) tasks.

3.2 The ProPublica report

The criminal justice system across the globe has started using algorithms to assess if an individual who had committed some crime in the past (aka defendant) is going to re-offend in the future (aka recidivism). In a classic report in 2016 published in ProPublica, a group of journalists showed how a popular software COMPAS (Correctional Offender Management Profiling for Alternative Sanctions) released by Northpointe, Inc. was severely biased against black defendants.

3.2.1 Data gathered by the authors

Defendant scoring. The authors collected a dataset of two years of COMPAS scores from Broward County Sheriff's office in Florida through a public records request. This score is primarily used to identify if a defendant should be detained or released before the trial. This resulted in a dataset of 11 757 individuals who were assessed at the pretrial stage spanning the years 2013 and 2014. A score on the scale of 1–10 was assigned to each of the defendants where one indicates least risk and ten indicates highest risk. The COMPAS scores are interpreted as follows—1–4: 'low', 5–7:

'medium', and 8–10: 'high.' The above score is assigned for a defendant in at least three categories—'risk of recidivism,' 'risk of violence,' and 'risk of failure to appear.'

Defendant profiling. The authors created a profile of each defendant in the COMPAS database, based on their criminal records. The criminal records were obtained from Broward County Clerk's Office website. The defendants from the COMPAS database were mapped to their criminal records through their first name, last name, and date of birth. The profile of each person was built two times—once before and once after they were scored using COMPAS. In order to determine the race, the authors used the Broward County Sheriff's Office classification format that identifies defendants belonging to one of the following races: black, white, Hispanic, Asian and Native American, and other. In addition, the authors collected the imprisonment records of the defendants from the Broward County Sheriff's Office from January 2013 to April 2016, and public incarceration records from the Florida Department of Corrections website.

3.2.2 Defining recidivism

The authors defined recidivism as the conduct of a criminal offense resulting in a jail booking which took place after the crime for which the defendant was originally COMPAS scored. One issue in this whole process is to correctly connect the criminal case to the defendant's COMPAS score. In order to have this connection the authors resorted to the following heuristic—cases with arrest or charge dates within 30 days of the COMPAS screening were connected to the COMPAS score. In some instances the COMPAS score could not be connected to any charge and these were removed by the authors from their analysis. As a next step the authors determined if a defendant had been charged with a new crime after the crime for which they had been COMPAS screened. The authors set a period of two years within which the new crime should have been committed based on Northpointe's practitioners' guide.

3.2.3 Primary observations

The main objective of the authors was to test the existence of racial disparities (if any) in the COMPAS scores. In order verify this they designed a simple logistic regression model using the following features—race, age, criminal history, future recidivism, charge degree, gender, and age to predict the COMPAS score (low versus medium and high score ranges). Some of the most predictive factors in the model are noted in table 3.1.

Among all the features, the authors found that the age of a defendant had the highest predictive capability. Surprisingly, younger defendants with age less than 25 were 2.5 times more likely to be assigned a higher COMPAS score than middle-aged offenders (i.e., with age greater than 45). This result was obtained while controlling for other confounding factors like prior crimes, future criminality, race, and gender.

A second and an even more surprising observation was that certain races were quite predictive of a higher score. Black defendants in particular had overall higher

Table 3.1. Predictive capability of the different features. *:$p < 0.1$, **:$p < 0.05$, ***:$p < 0.01$.

Features	Predictive capability
Female	0.221***
Age: greater than 45	−1.356***
Age: less than 25	1.308***
Black	0.477***
Asian	−0.254
Hispanic	−0.428***
Native American	1.394*
Other	−0.826***
Number of priors	0.269***
Misdemeanor	−0.311***
Two year recidivism	0.686***

Table 3.2. L-R: contingency table for all defendants (**FPR: 32.35%, FNR: 37.40%**), contingency table for black defendants (**FPR: 44.85%, FNR: 27.99%**), contingency table for white defendants (**FPR: 23.45%, FNR: 47.72%**). Sur.: survived, Rec.: recidivated.

All	Low	High
Sur.	2681	1282
Rec.	1216	2035
Black	**Low**	**High**
Sur.	990	805
Rec.	532	1369
White	**Low**	**High**
Sur.	1139	349
Rec.	461	505

scores of recidivism. Even after controlling for all other factors, black individuals were found to be 45% more likely to be assigned a higher score than whites.

The authors also found that female defendants were 19.4% more likely to be assigned higher score than males despite the fact that females had a low overall level of criminality.

As a next step the authors studied the error rates. For this purpose once again they assumed defendants with COMPAS score other than 'low' to be at the risk of recidivism. This was compared against whether they actually recidivated in the following two years. The contingency tables for all, black and white defendants are presented in table 3.2. The table shows that the black defendants who did not recidivate in future were at least two times more likely to be misclassified by COMPAS to be at higher risk compared to white defendants. On the other hand COMPAS made the opposite mistakes with the white defendants. It under-classified white re-offenders as being at low risk 70.5% times more often than black re-offenders. This along with several other nuanced analyses led the authors to conclude that the COMPAS software is severely unfair to the black defendants and is therefore highly inappropriate for use in criminal justice systems.

In a follow up study [1] the authors showed that COMPAS is no more accurate or fair than individuals with no or very little knowledge in criminal justice. They further showed that a simple logistic regression based classifier with just two features (age and the number of past convictions) performs as well as the proprietary (black box) COMPAS algorithm which known to be based on 137 features.

3.3 Gender shades

In this classic work the authors investigated the biases present in commercial facial recognition systems in terms of their gender detection performance. The authors also identified biases in benchmark datasets with respect to phenotypic subgroups. The authors resorted to the dermatologist approved Fitzpatrick Skin Type classification system[1], to characterize the gender and skin type distribution. A gender (male/female) and a skin type (lighter/darker) pairing is referred to as an intersectional group.

3.3.1 Biases in existing gender classification benchmark datasets

The two benchmark datasets that the authors investigated were IJB-A and Adience.

IJB-A. This is a US government benchmark dataset that was released by the National Institute of Standards and Technology (NIST) in 2015. Since this dataset has the direct involvement of the government, the authors chose it as a benchmark. During the time of their experiments, the dataset contained images of 500 unique individuals who were public figures. One image of each of these individuals was annotated manually by one of the authors. Both the gender and the skin type were annotated. The authors found that in this dataset, 79.6% individuals were of lighter skin types. Darker females were the least represented making up only 4.4% of the dataset. Darker males were 16%, lighter females were 20.2% and lighter males comprised the largest portion of the dataset at 59.4%.

Adience. This benchmark was released in 2014 and contained 2284 unique individuals at the time of release. Out of these, 2014 images were discernible and hence manually annotated for the skin type based on the Fitzpatrick scale. The gender annotations were already present in this dataset. In this dataset the authors found 86.2% individuals were of lighter skin type. Darker males were only 6.4% of the whole dataset. Darker females were 7.4%, lighter males were 41.6% and lighter females had the largest representation of 44.6%.

---

[1] https://en.wikipedia.org/wiki/Fitzpatrick_scale

3.3.2 Construction of a new benchmark dataset

Due to the existing biases in the benchmark datasets, the authors introduced a new dataset that had a balanced representation of all the intersectional groups. They chose parliamentarians from three African (Rwanda, Senegal, South Africa) and three European countries (Iceland, Finland, Sweden). These countries were chosen based on their ranking as per gender parity released by the Inter Parliamentary Union. The dataset which the authors named Pilot Parliaments Benchmark (PPB) contains images of 1270 individuals. The gender annotation was done by three independent annotators and the skin type annotation was done by a registered dermatologist. This dataset contains 53.6% lighter skin individuals and 46.4% darker skin individuals. 21.3%, 25%, 23.3%, and 30.3% individuals are darker females, darker males, lighter females, and lighter males respectively.

3.3.3 Biases in gender detection

The authors used three commercial gender detection softwares namely the Microsoft Face API [2], the IBM Watson Visual Recognition API [3], and Face++ [4]. For the evaluation of these softwares the authors used the PPB benchmark since it has the most balanced representation of the intersectional groups. Table 3.3 shows the error rates, i.e., (1 − PPV) across the different intersectional groups and the three softwares.

Overall, as the table notes, the authors find that all the softwares perform the worst for the darker skinned females (error rates 20.8%–34.7%). All classifiers perform better on lighter faces than the darker faces (11.8%–19.2% difference in error rate). IBM and Microsoft performs best for lighter males and Face++ does best for darker males.

3.4 Fair ML

3.4.1 Fair classification

For the main discussions in this book we shall restrict ourselves to a binary classification framework and would refer to the work presented in [5]. The basic formulation is extendable to a m-ary classification framework.

As we have already discussed in the first chapter of this book, for a binary classification task we need to find a function $f(\mathbf{x})$ that maps a feature vector $\mathbf{x}$ to a binary label $y = \{-1, 1\}$. This is done using a set of N training data points which

Table 3.3. % Error rates across different intersectional groups and different commercial softwares. **DF:** darker female, **DM:** darker male, **LF:** lighter female, **LM:** lighter male.

Software	DF	DM	LF	LM
Microsoft	**20.8**	6.0	1.7	0.0
Face++	**34.5**	0.7	6.0	0.8
IBM	**34.7**	12.0	7.1	0.3

can be represented as $\{(\mathbf{x}_i, y_i)\}_{i=1}^N$. In case of margin based classifiers this is simply the decision boundary expressed as a set of parameters denoted by θ. The objective is to find the parameters θ^* that correspond to decision boundary resulting in the best classification accuracy on the unseen test set. As discussed earlier this is usually achieved by minimizing a loss function L over the training set. More formally, $\theta^* = \text{argmin}_\theta L(\theta)$. Recall that $\{z_i\}_{i=1}^N$ are the sensitive attributes. In case one or more of these attributes are correlated with the training labels, it can adversely affect the performance of the classifier. For instance, if we assume d_{θ^*} to be the signed distance of the vector $\mathbf{x}$ from the decision boundary and $f_\theta(\mathbf{x}_i) = 1$ if $d_{\theta^*}(\mathbf{x}_i) \geq 0$, then the percentage of users in the protected class with $d_{\theta^*}(\mathbf{x}_i) \geq 0$ might drastically differ from the other class.

Compliance with the disparate treatment. In order to comply with disparate treatment, one has to ensure that the sensitive attributes are not a part of the feature set used for the purpose of prediction. Hence we have $\{x_i\}_{i=1}^N$ and $\{z_i\}_{i=1}^N$ as disjoint.

Compliance with disparate impact. To comply with disparate impact, one of the most well-known criteria has been to resort to the $p\%$-rule. This rule states that the ratio of the percentage of individuals from the protected class with $d_{\theta^*}(\mathbf{x}) \geq 0$ to that of the other class should be no less than $p: 100$. Formally, and without any loss of generality, if we assume that there is only one binary sensitive attribute $z = \{0, 1\}$, then we have

$$\min\left(\frac{P(d_\theta(x) \geq 0|z=1)}{P(d_\theta(x) \geq 0|z=0)}, \frac{P(d_\theta(x) \geq 0|z=0)}{P(d_\theta(x) \geq 0|z=1)}\right) \geq \frac{p}{100} \quad (3.1)$$

However, the above constraint cannot be imposed on standard decision boundary estimation algorithms because of its non-convex nature. Hence an alternative approximation as follows was introduced by the authors as a proxy for the $p\%$-rule.

The authors in [5] noted that one way to measure the (un)fairness should be to estimate the covariance between the users' sensitive attributes, $\{\mathbf{z}_i\}_{i=1}^N$, and the signed distance from the users' feature vectors to the decision boundary, $\{d_\theta(\mathbf{x}_i)\}_{i=1}^N$. Formally,

$$\text{Cov}(\mathbf{z}, d_\theta(\mathbf{x})) = \mathbb{E}[(\mathbf{z} - \bar{\mathbf{z}})d_\theta(\mathbf{x})] - \mathbb{E}[(\mathbf{z} - \bar{\mathbf{z}})]\bar{d}_\theta(\mathbf{x})$$
$$\approx \frac{1}{N}\sum_{i=1}^N (\mathbf{z}_i - \bar{\mathbf{z}})d_\theta(\mathbf{x}_i) \quad (3.2)$$

Note that here the term $\mathbb{E}[(\mathbf{z} - \bar{\mathbf{z}})]\bar{d}_\theta(\mathbf{x})$ vanishes since $\mathbb{E}[(\mathbf{z} - \bar{\mathbf{z}})]$ is zero. Generally in linear classifiers like SVMs and logistic regression the decision boundary is the hyperplane defined by $\theta^T \mathbf{x} = 0$. Thus one can suitably replace $d_\theta(\mathbf{x})$ in equation (3.2) to obtain

$$\text{Cov}(\mathbf{z}, d_\theta(\mathbf{x})) = \sum_{i=1}^N (\mathbf{z}_i - \bar{\mathbf{z}})\theta^T \mathbf{x}_i \quad (3.3)$$

Having this covariance setup has two advantages. First, note that equation (3.3) is convex as opposed to the $p\%$-rule constraint (equation 3.1). Second, if p is maintained at 100, the covariance should approach zero.

Maximizing accuracy under fairness constraints: now if one has to maximize accuracy under fairness constraints, one would have to minimize the following loss function over the training set to obtain an estimate of θ. Formally,

$$\text{minimize } L(\theta)$$
$$\text{subject to } \frac{1}{N}\sum_{i=1}^{N}(\mathbf{z}_i - \bar{\mathbf{z}})\theta^T \mathbf{x}_i \leq \mathbf{c}, \quad (3.4)$$
$$\frac{1}{N}\sum_{i=1}^{N}(\mathbf{z}_i - \bar{\mathbf{z}})\theta^T \mathbf{x}_i \geq -\mathbf{c}$$

Here $\mathbf{c}$ is a threshold that bounds the covariance of the sensitive attribute and the signed distance between the feature vector and the decision boundary. As, $\mathbf{c} \to 0$, the classifier tends to be fairer.

Maximizing fairness under accuracy constraints: in certain scenarios, the correlation between the class labels and the sensitive attribute(s) could be very high. In such cases, having a fairness constraint imposed might adversely affect the accuracy of the model. A suitable alternative in such cases is to maximize fairness (minimize disparate impact) with accuracy being the constraint. The revised formulation in this case would be

$$\text{minimize } \left| \frac{1}{N}\sum_{i=1}^{N}(\mathbf{z}_i - \bar{\mathbf{z}})\theta^T \mathbf{x}_i \right| \quad (3.5)$$
$$\text{subject to } L(\theta) \leq (1 + \gamma)L(\theta^*)$$

In this equation, $L(\theta^*)$ is the optimal loss obtained by training the unconstrained classifier. $\gamma \geq 0$ implies the additional allowable loss over that of the unconstrained classifier. If $\gamma \to 0$, then the optimal loss of the unconstrained classifier should be (almost) equal to that of the constrained classifier thus trying to reach the maximum achievable accuracy.

Choice of the loss function. The choice of the loss function $L(\theta)$ is determined by the classifiers used for a task. For instance, if we have a logistic regression then the loss function is given by the maximum likelihood

$$L(\theta) = -\sum_{i=1}^{N} \log p(y_i | \mathbf{x}_i, \theta) \quad (3.6)$$

and the equation (3.4) would have to be re-written as

$$\text{minimize } -\sum_{i=1}^{N} \log p(y_i | \mathbf{x}_i, \theta)$$
$$\text{subject to } \frac{1}{N}\sum_{i=1}^{N}(\mathbf{z}_i - \bar{\mathbf{z}})\theta^T \mathbf{x}_i \leq \mathbf{c}, \quad (3.7)$$
$$\frac{1}{N}\sum_{i=1}^{N}(\mathbf{z}_i - \bar{\mathbf{z}})\theta^T \mathbf{x}_i \geq -\mathbf{c}$$

If the classifier at hand is a linear SVM, then the loss function would look like

$$L(\boldsymbol{\theta}) =$$

$$\text{minimize } ||\boldsymbol{\theta}||^2 + C\sum_{i=1}^{N}\xi_i \quad (3.8)$$

$$\text{subject to } y_i\boldsymbol{\theta}^T\mathbf{x}_i \geq 1 - \xi_i, \forall i \in \{1,\ldots,N\}$$

$$\xi_i > 0, \forall i \in \{1,\ldots,N\}$$

where $||\boldsymbol{\theta}||^2$ refers to the margin between the support vectors assigned to the two binary classes and $C\sum_{i=1}^{N}\xi_i$ is an adjustment to penalize the data points falling inside the margin. Therefore in this case the revised form of the equation (3.4) would be

$$\text{minimize } ||\boldsymbol{\theta}||^2 + C\sum_{i=1}^{N}\xi_i$$

$$\text{subject to } y_i\boldsymbol{\theta}^T\mathbf{x}_i \geq 1 - \xi_i, \forall i \in \{1,\ldots,N\}$$

$$\xi_i \geq 0, \forall i \in \{1,\ldots,N\} \quad (3.9)$$

$$\frac{1}{N}\sum_{i=1}^{N}(\mathbf{z}_i - \bar{\mathbf{z}})\boldsymbol{\theta}^T\mathbf{x}_i \leq \mathbf{c},$$

$$\frac{1}{N}\sum_{i=1}^{N}(\mathbf{z}_i - \bar{\mathbf{z}})\boldsymbol{\theta}^T\mathbf{x}_i \geq -\mathbf{c}$$

The above constrained optimization can be solved for $\boldsymbol{\theta}$ by writing a quadratic program.

The idea can also be extended to non-linear SVMs. The trick is to transform the feature vectors $\mathbf{x}$ to a higher dimension using some operator $\Phi(\cdot)$. The decision boundary equation therefore is $\boldsymbol{\theta}^T\Phi(\mathbf{x}) = 0$. Here one needs to resort to the dual form of the problem since the original form might have a very complex feature space due to the transformation thus making it difficult to solve using a quadratic program. In this case the revised equation (3.4) would take the form

$$\text{minimize } \sum_{i=1}^{N}\alpha_i + \sum_{i=1}^{N}\alpha_i y_i(g_\alpha(\mathbf{x}_i) + h_\alpha(\mathbf{x}_i))$$

$$\text{subject to } \alpha_i \geq 0, \forall i \in \{1,\ldots,N\},$$

$$\sum_{i=1}^{N}\alpha_i y_i = 0, \quad (3.10)$$

$$\frac{1}{N}\sum_{i=1}^{N}(\mathbf{z}_i - \bar{\mathbf{z}})g_\alpha(\mathbf{x}_i) \leq \mathbf{c},$$

$$\frac{1}{N}\sum_{i=1}^{N}(\mathbf{z}_i - \bar{\mathbf{z}})g_\alpha(\mathbf{x}_i) \geq -\mathbf{c}$$

where α are the dual variables replacing θ. The function $g_\alpha(\mathbf{x}_i) = \sum_{j=1}^{N}\alpha_j y_j k(\mathbf{x}_i, \mathbf{x}_j)$ is the newly signed distance from the decision boundary in the transformed space. The function $h_\alpha(\mathbf{x}_i) = \sum_{j=1}^{N}\alpha_j y_j \frac{1}{C}\delta_{ij}$ where $\delta_{ij} = 1$ if $i = j$ and 0 otherwise. The operation $k(\mathbf{x}_i, \mathbf{x}_j) = \langle \Phi(\mathbf{x}_i), \Phi(\mathbf{x}_j) \rangle$ denotes the *kernel* or the inner product between a pair of transformed feature vectors.

Key results. The authors in [5] performed a series of rigorous experiments to establish the effectiveness of the algorithms developed. The experiments were done on both synthetic as well as real-world datasets. In the synthetic experiments the authors generated a bunch of binary class labels uniformly at random and associated each label with a 2-dimensional feature vector. The feature vectors were constructed by drawing samples from bivariate Gaussian distributions. Two such Gaussian distributions (sufficiently different in their means and variances) were used to draw feature vectors for the positive and negative labels respectively. Next they constructed the sensitive attribute from a Bernoulli distribution of the form $p(z = 1) = p(\mathbf{x}'|y = 1)/(p(\mathbf{x}'|y = 1) + p(\mathbf{x}'|y = -1))$, where $\mathbf{x}' = [\cos(\phi), -\sin(\phi); \sin(\phi), \cos(\phi)]\mathbf{x}$ is simply a rotated version of the feature vector, $\mathbf{x}$. The value of ϕ here determines the extent of the correlation of the sensitive attribute z with the feature vector $\mathbf{x}$. Applying their fairness aware algorithm on this dataset the authors showed how the learnt decision boundary itself gets rotated (with respect to the decision boundary of the unconstrained classifier) to make the classification results fair. The authors also reported the performance of their model on two real datasets —adult income dataset [6] and bank marketing dataset [7]. These datasets had different sensitive attributes including gender, race, and age. The authors showed through a rigorous set of experiments that their fairness aware algorithms were able to make largely unbiased predictions without hurting the performance of the system much. I would refer the reader to the source paper [5] for a more detailed account of the results.

3.4.2 Fair regression

Designing classifiers are useful when the decision space is small and discrete like accept/reject applications for job hiring and school/college admission. However, in many applications the decision space is continuous and the prediction therefore has to be real-valued. For instance, one might like to make predictions about the GPA/rank of first year college students or the overall success of a person in a job. In a semi-automated workflow, the human decision makers might require to observe prediction scores rather than simplistic yes/no answers to reach a final decision (see [8–10]). A fair classification tool can always be used to make the final decision in such cases but this reduces the decision autonomy of the human decision makers and is, therefore, often disfavored. Ranking or scoring algorithms do not suffer from this drawback but need to be adjusted to include the necessary fairness constraints.

In [11], the authors attempt to propose a solution to this problem. Like in the case of classification, here also we need to find a function $f(\mathbf{x})$ that maps a feature vector $\mathbf{x}$ to a continuous label $y \in [0, 1]$. This is done using a set of N training data points which can be represented as $\{(\mathbf{x}_i, y_i)\}_{i=1}^{N}$ with the only exception that all y_i are real-valued and continuous (i.e., $y_i \in [0, 1]$). The accuracy of a prediction by $f(\mathbf{x}_i)$ on a model y_i is adjudged by a suitable loss function L. For instance the prediction of GPA for first year college students can be modeled as a regression problem where each y_i corresponds to a GPA value that is suitably normalized between $\in [0, 1]$. The error or the loss function could be the squared loss, i.e., $L(y_i, f(\mathbf{x}_i)) = \frac{(y_i - f(\mathbf{x}_i))^2}{2}$.

Now let us assume that we wish to ensure statistical parity. Since $f(\mathbf{x}_i) \in [0, 1]$ this boils down to having $P[f(\mathbf{x}_i) \geq s | z = 1] = P[f(\mathbf{x}_i) \geq s]$ where $s \in [0, 1]$ and z is as usual the protectedattribute. We then need to have a fairness accuracy knob like the parameter **c** used in equation (3.4). Thus the objective function takes the following form.

$$\min_{f \in F} \mathbb{E}[L(y_i, f(\mathbf{x}_i))]$$
subject to (3.11)
$$| P[f(\mathbf{x}_i) \geq s | z = 1] - P[f(\mathbf{x}_i) \geq s] | \leq c.$$

where F is the family of functions from which f is to be searched. Different values of parameter **c** can be chosen to vary the strength of the constraint in favor of the protected group.

The authors then solved the above fair regression problem using a cost-sensitive classification (CS) oracle. They recast the problem into a constrained and cost-sensitive classification problem and solved it via a reduction technique proposed in one of their own earlier works [12] by repeatedly invoking the CS oracle. The authors used three datasets to evaluate their method. These are the adult income dataset [6], the law school dataset [13], and the communities and crime dataset [14]. For all the datasets the authors showed that their algorithm achieves statistical parity without hurting the accuracy much. I would refer the reader to [11] for a more a detailed account of the reduction based algorithm and the evaluation.

3.4.3 Fair clustering
Different fairness notions for clustering problems exist in the literature. In the following, we discuss the most popular ones.

3.4.3.1 Balance
This is a group-level notion of fairness that was proposed by Chierichetti *et al* [15] for a setting with two protected groups and later generalized by Bera *et al* [16] for multiple protected groups. It is one of the most popular fairness metrics used in fair clustering [17–19]. Let us assume that there are N protected groups and K different clusters. Further, let ρ and ρ_C be respectively the proportion of samples of the dataset belonging to the protected group z and the proportion of samples in cluster $C \in [K]$ belonging to protected group z. Next, they defined another ratio for this cluster and the protected group as $\rho_{C,z} = \frac{\rho}{\rho_C}$. Based on this the balance can be formulated over all the clusters and the protected groups as follows.

$$\min_{C\in[K], z\in[N]} \min\left(\rho_{C,z}, \rho_{C,z}^{-1}\right) \tag{3.12}$$

Balance, therefore, lies between [0, 1] and a higher value of balance indicates a more fair clustering output. Hence a fair clustering algorithm shall attempt to maximize the balance metric to improve fairness.

3.4.3.2 Social fairness
The idea of social fairness was proposed for the K-means clustering objective [20]. This is again a group-level fairness notion. The cost function of the K-means algorithm is formulated as follows. Let $\mathbf{U}$ be the cluster centers for the K different clusters. For the input dataset $\mathbf{x}$, let us define the cost of the K-means clustering as follows.

$$O(\mathbf{U}, \mathbf{x}) = \sum_{\mathbf{x}_i \in \mathbf{x}} \min_{\mathbf{U}_i \in \mathbf{U}} \|\mathbf{x}_i - \mathbf{U}_i\|^2 \tag{3.13}$$

Let $\mathbf{x}_z$ denote the samples of $\mathbf{x}$ that belong to the protected group z. The social fairness is then defined as follows.

$$\max_{z\in[N]} \frac{O(\mathbf{U}, \mathbf{x}_z)}{|\mathbf{x}_z|} \tag{3.14}$$

Unlike balance which needs to be maximized, social fairness needs to be minimized since it is a distance function.

3.4.3.3 Bounded representation
The notion of bounded fairness [21] is defined in terms of two parameters α (upper bound) and β (lower bound) and works at the group-level. Let us assume $\rho_{C,z}$ be the proportion of protected group members $z \in [N]$ in the cluster $C \in [K]$. The (α, β)-bounded representation requires that

$$\beta \leqslant \rho_{C,z} \leqslant \alpha, \forall C \in [K], z \in [N] \tag{3.15}$$

Note that this is a constraint based definition and the clustering is fair if the constraints for each group and each cluster are maintained.

3.4.3.4 Distributional fairness
This individual-level fairness notion was first introduced in [22]. One assumes that a fairness similarity notion $S \in \mathbb{R}^+$ is known to exist between the pair of samples in the dataset $\mathbf{X}$. The idea is that the distance obtained using KL-divergence (or any other f-divergence for that matter [23]) for the output distributions of each sample pair should be smaller than the S-metric based distance. Further one also assumes the knowledge of the cluster centers.

Let $\mathbf{U}$ be the cluster centers for the K different clusters. Further let the KL-divergence between the distributions $D_\mathbf{x}$ and $D_\mathbf{y}$ for the pair of samples $\mathbf{x}, \mathbf{y} \in \mathbf{X} \times \mathbf{X}$ cast over $\mathbf{U}$ be denoted as $KL(D_\mathbf{x} \| D_\mathbf{y})$. The distributional fairness requires that

$$KL(D_\mathbf{x}\|D_\mathbf{y}) \leqslant S(\mathbf{x}, \mathbf{y}) \qquad (3.16)$$

for all pairs of data samples $\mathbf{x}, \mathbf{y} \in \mathbf{X} \times \mathbf{X}$.

3.4.3.5 Individual fairness
This notion of individual fairness was proposed by Kleindessner et al [24]. Let the output of a clustering algorithm be denoted by the set $\{C_1, C_2, \ldots, C_K\}$. For each data point $\mathbf{x} \in \mathbf{X}$, let d be a well-defined clustering distance metric and $\mathbb{C}_a$ be the cluster to which $\mathbf{x}$ belongs to. The concept of fairness is defined as a set constraints for the data point $\mathbf{x}$ and all clusters $b \in [K]$, $b \neq a$ as follows.

$$\frac{1}{|\mathbb{C}_a| - 1} \sum_{\mathbf{p} \in C_a} d(\mathbf{x}, \mathbf{p}) \leqslant \frac{1}{|\mathbb{C}_b|} \sum_{\mathbf{p} \in C_b} d(\mathbf{x}, \mathbf{p}) \qquad (3.17)$$

The clustering is said to be individually fair if all the individual samples in the dataset $\mathbf{X}$ obey the above constraint.

3.4.4 Fair embedding

Vector representation of words (aka word embeddings) is ubiquitous in NLP and IR applications. The basic idea is to represent each word w as a d-dimensional vector, i.e., $w \in \mathbb{R}^d$. The word vectors are trained on large text corpora using standard skip-gram [25] or continuous-bag-of-words models [25]. The words that have similar semantics should be 'closer' to one another in the embedding space. These vectors have been found to be very effective in solving *word analogies* such as 'man is to king as woman is to x'. A simple vector arithmetic of the form

$$\overrightarrow{\text{man}} - \overrightarrow{\text{woman}} = \overrightarrow{\text{king}} - \vec{x} \qquad (3.18)$$

tells that the best solution is when $x \approx \overrightarrow{\text{queen}}$. Solving such analogy tasks are the building blocks of numerous NLP applications sentiment analysis, document understanding, etc. While this has brought in a lot of new hope in the scientific community, certain pitfalls have also been identified. One of the negative consequences that have become quite concerning is that such embeddings have notably been found to be sexist. For instance, if one attempts to solve the following analogy puzzle

$$\overrightarrow{\text{man}} - \overrightarrow{\text{woman}} = \overrightarrow{\text{doctor}} - \vec{x} \qquad (3.19)$$

the best solution that one obtains is $x \approx \overrightarrow{\text{nurse}}$. The problem here is that although the two occupation words 'doctor' and 'nurse' are gender neutral, they get associated with 'man' and 'woman' respectively due to the biases in the text from which the embeddings are learnt. Therefore there is a dire need to intervene in the construction of such embeddings and debisas them from being sexist. While there have been many works in this direction, we shall discuss in brief the most popular and the first of its kind study described in [26].

Preliminaries. Let W be the set of all words. All the word embeddings ($\vec{w} \in W$) are represented as unit vectors in $\mathbb{R}^d$, i.e., $\|\vec{w}\| = 1$. Further $N \subset W$ is the set of gender neutral words like 'receptionist', 'architect', etc. Similarly there is a set $D \subset W \times W$ of gender defining F–M pairs such as {'she', 'he'}, {'woman', 'man'}, {'gal', 'guy'} etc. The authors in [26] detail an elaborate strategy to obtain these two sets. The similarity between a pair of word vectors (read words) is the inner product of the vectors, i.e., $\vec{w_1} \cdot \vec{w_2}$. Since the word vectors are normalized so $\cos(\vec{w_1}, \vec{w_2}) = \frac{\vec{w_1} \cdot \vec{w_2}}{\|\vec{w_1}\| \|\vec{w_2}\|} = \vec{w_1} \cdot \vec{w_2}$.

Debiasing algorithm. The authors described the algorithm in three subparts—(i) *identification of the gender subspace*, (ii) *neutralize*, and (iii) *equalize*. In the first step the idea is to obtain a common direction/subspace that captures the overall gender component in the embeddings. This then would be useful to debias the gender neutral words. The neutralize step ensures that gender neutral words become zero in the gender subspace. The equalize step attempts to make sets of words outside the subspace to be equidistant from the gender neutral words. Thus the word 'nanny' should be at an equal distance from the equality set {female, male}. We shall now see how to mathematically formulate these steps.

Identification of the gender subspace: a subspace B is a set of k orthogonal unit vectors $\{\vec{b_1}, \vec{b_2}, \ldots, \vec{b_k}\} \subset \mathbb{R}^d$. If $k = 1$, then the subspace reduces to a single direction. The projection of a vector $\vec{w}$ on B is defined as

$$\vec{w_B} = \sum_{j=1}^{k} (\vec{w} \cdot \vec{b_j}) \vec{b_j} \tag{3.20}$$

In order to compute the *gender subspace*, we need the defining sets $D_1, D_2, D_3 \ldots D_n \subset D$, the value of the integer k and the word embeddings $\vec{w} \in \mathbb{R}^d$. We first obtain the means of the defining sets as

$$\vec{\mu_i} = \sum_{w \in D_i} \frac{\vec{w}}{D_i} \tag{3.21}$$

The mean shift of the individual word vectors can be obtained as $\vec{w} - \vec{\mu_i}$. This shifts all the vectors to the origin and makes processing easier. Note that we are interested in obtaining that direction or subspace which explains the maximum variance of the space. This can be obtained by extracting the principal singular vectors of the covariance matrix **C** defined as follows.

$$\mathbf{C} = \sum_{i=1}^{n} \sum_{\vec{w} \in D_i} \frac{(\vec{w} - \vec{\mu_i})^T (\vec{w} - \vec{\mu_i})}{D_i} \tag{3.22}$$

The gender subspace B is the first k rows of **SVD**(C).

Neutralize: for each word $w \in N$ in the set of gender neutral words (which could be potentially biased), it is re-embedded as follows.

$$\vec{w} \Longleftarrow \frac{(\vec{w} - \overrightarrow{w_B})}{||(\vec{w} - \overrightarrow{w_B})||} \tag{3.23}$$

where w_B is the projection of w on the gender subspace B. This essentially removes the (potential) gender bias present in the embedding w. Figures 3.1(a) and (b) show an example of how the embeddings of the gender neutral words 'nurse' and 'doctor' look before and after the *neutralize* step respectively.

Equalize: Let the family of equality sets be represented by $\Xi = \{E_1, E_2, \ldots, E_m\}$ where $E_i \in W$. For each equality set $E_i \in \Xi$ let

$$\vec{\mu} = \sum_{\vec{w} \in E} \frac{\vec{w}}{E} \tag{3.24}$$

Every word $\vec{w} \in E$ can be represented as the sum of the projection in the gender subspace, i.e., $\overrightarrow{w_B}$ and the component outside this subspace, i.e., $\overrightarrow{w_\perp}$. Within the bias subspace we move everything to mean 0, i.e., replace $\overrightarrow{w_B}$ with $\overrightarrow{w_B} - \overrightarrow{\mu_B}$. Further let us call $\overrightarrow{w_\perp}$ as $\vec{\nu}$. Thus, we have $\vec{w} = \vec{\nu} + (\overrightarrow{w_B} - \overrightarrow{\mu_B})$. Since we are operating with unit vectors we would like to keep $||w||^2 = 1$. Thus, $||\vec{\nu} + (\overrightarrow{w_B} - \overrightarrow{\mu_B})||^2 = 1$ or $||\overrightarrow{w_B} - \overrightarrow{\mu_B}|| = \sqrt{1 - ||\vec{\nu}||^2}$. We can therefore write

$$\vec{w} = \vec{\nu} + \sqrt{1 - ||\vec{\nu}||^2} \frac{(\overrightarrow{w_B} - \overrightarrow{\mu_B})}{||\overrightarrow{w_B} - \overrightarrow{\mu_B}||} \tag{3.25}$$

Note that for the second term, one actually needs to compute $(\overrightarrow{w_B} - \overrightarrow{\mu_B})$; however, in the gender subspace that we have constructed, everything is in the form of unit vectors and therefore we can only have access to the term $\frac{(\overrightarrow{w_B} - \overrightarrow{\mu_B})}{||\overrightarrow{w_B} - \overrightarrow{\mu_B}||}$ which we multiply by $\sqrt{1 - ||\vec{\nu}||^2}$ to recover the actual $(\overrightarrow{w_B} - \overrightarrow{\mu_B})$. The figures 3.2 shows the

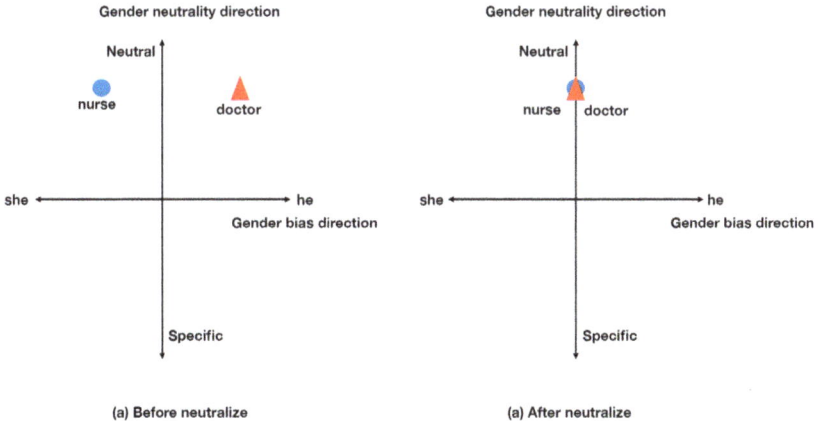

Figure 3.1. The embeddings of the gender neutral words 'nurse' and 'doctor' (a) before and (b) after the *neutralize* step respectively.

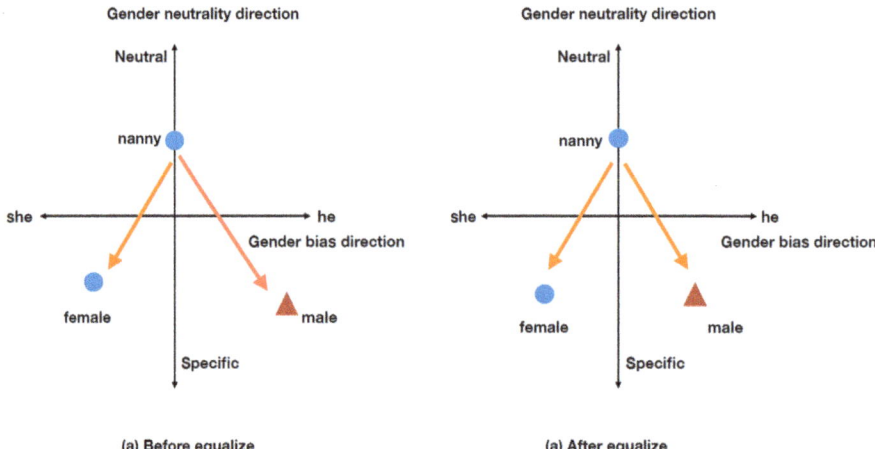

Figure 3.2. The embeddings of the gender words 'female' and 'male' (a) before and (b) after the *equalize* step.

embeddings of the gender words 'female' and 'male' before and after the *equalize* step respectively.

Evaluation metrics. The authors used two different metrics to evaluate the efficacy of the debiasing algorithm. Specifically, they considered two metrics—direct bias and indirect bias.

Direct bias: the authors quantify direct bias by formulating an analogy generation task. For instance, given a seed pair (*she, he*), the task would be to generate all pairs (x, y) such that *she* to x and *he* to y is a 'good' analogy. For a seed pair of words (a, b) the seed direction is obtained as $\vec{a} - \vec{b}$ which is the normalized difference between the two seed words. The authors consider (*she, he*) as the seed pair. The rest of the word pairs (x, y) are scored based on the following metric.

$$S_{(a,b)}(x, y) = \begin{cases} \cos(\vec{a}-\vec{b}, \vec{x}-\vec{y}) & \text{if } \|\vec{x}-\vec{y}\| \leq \delta \\ 0 & \text{otherwise} \end{cases} \quad (3.26)$$

where δ denotes a similarity threshold. There are two components that the metric attempts to take care of—while the selected analogy pair should be (almost) parallel to the seed direction, the individual words of the pair should also be semantically close to one another. For their experiments, the authors set $\delta = 1$ and do not output multiple analogies with the same x to reduce redundancy.

Indirect bias: the direct bias discussed above is a result of the direct similarities between gender-specific and gender neutral words. However, biases can also manifest due to certain gendered associations with gender neutral words. To identify such *indirect* biases, the authors took the gender neutral pair 'softball' and 'football' and projected all the occupation words on the $\overrightarrow{\text{softball}} - \overrightarrow{\text{football}}$ direction. They found that words like 'bookkeeper' and 'receptionist' are closer to 'softball' than 'football'. This results due to the indirect female associations of 'softball', 'bookkeeper', and 'receptionist'. To quantify indirect bias the authors decompose the word vector $\vec{w}$ into two parts: $\vec{w} = \overrightarrow{w_B} + \overrightarrow{w_\perp}$ where $\overrightarrow{w_B}$ is the projection of $\vec{w}$ onto the bias subspace B. The authors define the similarity of two words $\vec{w}$ and $\vec{v}$ contributed by the gender component as

$$\beta(\vec{w}, \vec{v}) = \frac{\vec{w} \cdot \vec{v} - \frac{\overrightarrow{w_\perp} \cdot \overrightarrow{v_\perp}}{\|\overrightarrow{w_\perp}\|_2 \|\overrightarrow{v_\perp}\|_2}}{\vec{w} \cdot \vec{v}} \qquad (3.27)$$

Thus if $\overrightarrow{w_B} = 0 = \overrightarrow{v_B}$ then $\beta(\vec{w}, \vec{v}) = 0$ and if $\overrightarrow{w_\perp} = 0 = \overrightarrow{v_\perp}$ then $\beta(\vec{w}, \vec{v}) = 1$.

Key results. The authors ran experiments on the w2vNEWS word embeddings [26]. They performed three types of evaluation—(i) quality of the embeddings, (ii) direct bias, and (iii) indirect bias.

Quality of the embeddings: in order to test the quality of the debiased embeddings, the authors performed standard word similarity tasks. The authors showed that performance of these tasks before and after debiasing remains the same.

Direct bias: the authors used an analogy generation task to identify the effect of debiasing. They automatically generated pairs that are analogous to *she–he* and asked Amazon Mechanical Turk workers whether these pairs indicated gender stereotypes. For the initial w2vNEWS embeddings, out of the top 150 analogies, 19% pairs were judged to reflect gender stereotypes by a majority among ten workers. In contrast, for the debiased embeddings, only 6% pairs were found to reflect gender stereotypes as per the majority of turkers.

Indirect bias: the authors identified the words closest to $\overrightarrow{softball} - \overrightarrow{football}$ direction after debiasing. They observed that new words closest to this direction were *infielder* and *major leaguer* which are gender neutral.

Practice problems

Q1. A COMPAS recidivism prediction is simulated and the results for black and white individuals are noted in table 3.4. Calculate the FPR and FNR values for all defendants, black defendants, and white defendants. Based on your calculations, comment on the relative misclassification between the two racial groups (which one is at a higher risk of misclassification?). The phrases in the table mean the same as in the ProPublica report.

Table 3.4. Contingency table for a simulated COMPAS recidivism scenario.

	All defendants		Black defendants		White defendants	
	Low	High	Low	High	Low	High
Survived	5348	1372	1650	1132	1650	280
Recidivated	347	420	189	255	145	80

Q2. An adverse impact is a situation where the 80%-rule for avoiding disparate impact is not adhered to. Here we use the example of male versus female applicants for faculty hires. We have ten male applicants with five male hires. We also have 40 female applicants with five female hires. Is there an adverse impact? Can you mathematically say so?

Q3. The paper 'Man is to Computer Programmer as Woman is to Homemaker? Debiasing Word Embeddings' [26] defines gender bias of a word w by its projection on the 'gender direction': $\vec{w} \cdot (\overrightarrow{he} - \overrightarrow{she})$ assuming all vectors are normalized. The larger a word's projection is on $\overrightarrow{he} - \overrightarrow{she}$ the more biased it is (DIRECT BIAS). In the following set of questions, we explore the debiasing algorithms and their effects on the word embeddings.
  (a) Write the steps of the HARD-DEBIASING algorithm as described in the paper.
  (b) Consider the embedding space shown in figure 3.3 before executing the HARD-DEBIASING algorithm. Roughly draw the space along with the words after the execution of the HARD-DEBIASING algorithm. What is the main observation?
  (c) Next, let us compare the number of male-biased neighborhood words versus bias of a typical word. Let us plot this for the occupation words mentioned in the paper [26] as shown in figure 3.4. To find the number of male-biased neighborhoods, let us compute the k-nearest neighbor words using the cosine similarity (where k is 100) and find how many of them are male-biased using the direction of DIRECT BIAS vector. What can you observe from this plot and what does it signify?
  (d) What are the implications from the above plots as compared to the previous work [26] on debiasing word embeddings?

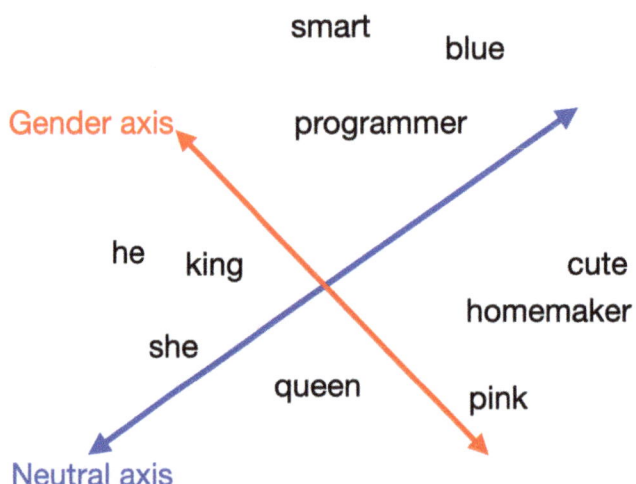

Figure 3.3. Embedding space before the execution of the HARD-DEBIASING algorithm.

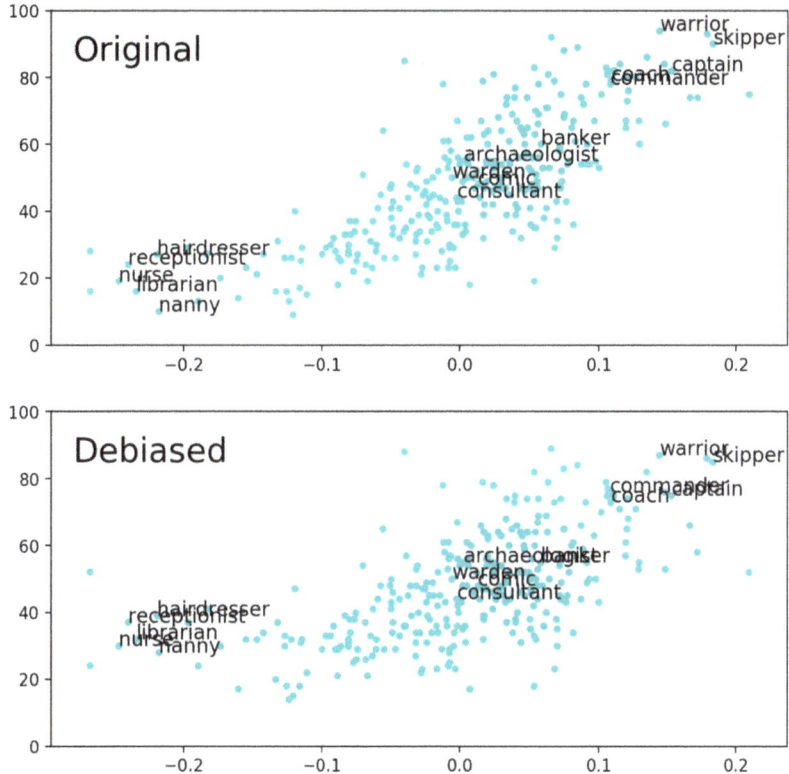

Figure 3.4. The plots for HARD-DEBIASED embedding, before (top) and after (bottom) debiasing. Reproduced from [27], CC BY 4.0.

References

[1] Dressel J and Farid H 2018 The accuracy, fairness, and limits of predicting recidivism *Sci. Adv.* **4** eaao5580

[2] Microsoft 2022 Microsoft Azure Face https://azure.microsoft.com/en-in/services/cognitive-services/face/

[3] IBM 2022 IBM Watson visual recognition, 2022. https://www.ibm.com/watson/services/visual-recognition/

[4] Face++ 2022 Face++ AI open platform https://www.faceplusplus.com/

[5] Zafar M B, Valera I, Rogriguez M G and Gummadi K P 2017 Fairness constraints: mechanisms for fair classification *Proc. 20th Int. Conf. on Artificial Intelligence and Statistics* vol 54; A Singh and J Zhu *Proc. Machine Learning Research* pp 962–70 https://jmlr.org/papers/v20/18-262.html

[6] Blake C L and Merz C J 1998 UCI repository of machine learning databases https://eric.ed.gov/?id=ED469370

[7] Moro S, Cortez P and Rita P 2014 A data-driven approach to predict the success of bank telemarketing *Decis. Support Syst.* **62** 22–31

[8] Waters A and Miikkulainen R 2014 GRADE: machine learning support for graduate admissions *AI Mag.* **35** 64

[9] US Federal Reserve 2007 Report to the congress on credit scoring and its effects on the availability and affordability of credit https://www.federalreserve.gov/boarddocs/rptcongress/creditscore/creditscore.pdf

[10] Lowenkamp C T, Johnson J L, Holsinger A M, VanBenschoten S W and Robinson C R 2013 The federal post-conviction risk assessment (PCRA): a construction and validation study *Psychol. Serv.* **10** 87–96

[11] Agarwal A, Dudík M and Wu Z S 2019 Fair regression: quantitative definitions and reduction-based algorithms *Proceedings of the 36th International Conference on Machine Learning (ICML 2019) (Long Beach, CA, 9–15 June 2019)*

[12] Agarwal A, Beygelzimer A, Dudik M, Langford J and Wallach H 2018 A reductions approach to fair classification *Proc. 35th Int. Conf. on Machine Learning* vol 80 ed J Dy and A Krause *Proc. Machine Learning Research* pp 60–9

[13] Wightman L F 1998 LSAC National Longitudinal Bar Passage Study *LSAC Research Report Series* (Newtown, PA: Law School Admission Council, Inc.) https://eric.ed.gov/?id=ED469370

[14] Redmond M and Baveja A 2002 A data-driven software tool for enabling cooperative information sharing among police departments *Eur. J. Oper. Res.* **141** 660–78

[15] Chierichetti F, Kumar R, Lattanzi S and Vassilvitskii S 2018 Fair clustering through fairlets *NIPS* https://dl.acm.org/doi/abs/10.5555/3295222.3295256

[16] Bera S, Chakrabarty D, Flores N and Negahbani M 2019 Fair algorithms for clustering *Fair algorithms for clustering Advances in Neural Information Processing* vol 32 ed H Wallach, H Larochelle, A Beygelzimer, F d'Alché-Buc, E Fox and R Garnett (Red Hook, NY: Curran Associates, Inc.) https://proceedings.neurips.cc/paper/2019/file/fc192b0c0d270dbf41870a63a8c76c2f-Paper.pdf

[17] Schmidt M, Schwiegelshohn C and Sohler C 2019 Fair coresets and streaming algorithms for fair *k*-means *Approximation and Online Algorithms: 17th Int. Workshop, WAOA 2019, Munich, Germany, September 12–13, 2019, Revised Selected Papers* (Berlin: Springer) pp 232–51

[18] Bercea I O, Khuller S, Rösner C, Schmidt M, Groß M, Kumar A and Schmidt D R 2019 On the cost of essentially fair clusterings *Approximation, Randomization, and Combinatorial Optimization. Algorithms and Techniques, APPROX/RANDOM, Leibniz Int. Proc. Informatics* ed D Achlioptas and L A Vegh (LIPIcs. Schloss Dagstuhl-Leibniz-Zentrum fur Informatik GmbH, Dagstuhl Publishing)

[19] Backurs A, Indyk P, Onak K, Schieber B, Vakilian A and Wagner T 2019 Scalable fair clustering *Proc. 36th Int. Conf. on Machine Learning* vol 97 ed K Chaudhuri and R Salakhutdinov *Proc. Machine Learning* pp 405–13 https://proceedings.mlr.press/v97/backurs19a.html

[20] Ghadiri M, Samadi S and Vempala S 2021 Socially fair k-means clustering *Proc. 2021 ACM Conf. on Fairness, Accountability, and Transparency* (New York: ACM) pp 438–48

[21] Ahmadian S, Epasto A, Kumar R and Mahdian M 2019 Clustering without over-representation *Proc. 25th ACM SIGKDD Int. Conf. on Knowledge Discovery and Data Mining, KDD'19* (New York: ACM) pp 267–75

[22] Anderson N, Bera S K, Das S and Liu Y 2020 *Distributional Individual Fairness in Clustering* (arXiv:2006.12589 [cs.LG])

[23] Ali S M and Silvey S D 1966 A general class of coefficients of divergence of one distribution from another *J. R. Stat. Soc. Ser. B* **28** 131–42

[24] Kleindessner M, Awasthi P and Morgenstern J 2020 A notion of individual fairness for clustering Proceedings of the 25th *International Conference on Artificial Intelligence and Statistics (AISTATS) 2022 (Valencia, Spain)* vol 151

[25] Mikolov T, Chen K, Corrado G and Dean J 2013 Efficient estimation of word representations in vector space *1st Int. Conf. on Learning Representations (ICLR) 2013* (Scottsdale, AZ, May 2–4, 2013) ed Y Bengio and Y LeCun

[26] Bolukbasi T, Chang K-W, Zou J Y, Saligrama V and Kalai A T 2016 Man is to computer programmer as woman is to homemaker? Debiasing word embeddings *30th Conf. on Neural Information Processing Systems (NIPS 2016) (Barcelona)*

[27] Gonen H and Goldberg Y 2019 Lipstick on a pig: Debiasing methods cover up systematic gender biases in word embeddings but do not remove them *Proc. 2019 Conf. of the North American Chapter of the Association for Computational Linguistics: Human Language Technologies, vol 1 (Long and Short Papers)* (Minneapolis, MN: Association for Computational Linguistics) pp 609–14

Chapter 4

Content governance

4.1 Chapter foreword

The definition of a 'platform' varies across contexts and over time. These contexts can range from forums to put forward your opinion to forums to facilitate commerce/transactions. Likewise, platforms can range from forums for public speech to online social networking websites; from bricks-and-mortar supermarkets to online e-commerce websites and so on. In this book, we limit our discussions to platforms as online websites and services that produce, host, organize, and distribute *content* among its stakeholders. The content could be of various types including chat messages, social media posts, multimedia, as well as items involving financial transactions such as products to be sold on e-commerce sites. However, some basic notions cut across these different definitions and applications of platforms. For example, the primary stakeholders of a platform are (a) the producer of content or item, (b) the consumer of content or item, and (c) the organization that runs the platform. For instance, on social networking sites, the users are both the producer and consumer of the content while the social networking company (e.g., Twitter, Facebook, Instagram, etc) is the owner of the platform. For an e-commerce platform, sellers are the item producers, customers/buyers are the item consumers and the e-commerce company (e.g., Amazon, Flipkart, Walmart, etc) is the owner of the platform that mediates between the seller and the customer. The mediation is primarily driven by algorithms which are most often proprietary to the company and, hence, inaccessible to scientists and practitioners for investigation. *Content governance* is a concept that refers to the layers of governance relationships that steer and regulate the interactions among various stakeholders including the producers, the consumers, the platform companies as well as the advertisers, the political actors, the policymakers, and governments. In this chapter, we shall consider two important aspects of this governance that have potential ethical ramifications. One of them deals with *content dissemination*, i.e., how the content is mediated among certain stakeholders of the platform and the biases that could potentially result from such dissemination. The second involves *content moderation*, i.e., detection and control of malicious, abusive, fake, and biased content.

4.2 Content dissemination

We shall take up two case studies to illustrate the potential ethical concerns due to content dissemination. The first one corresponds to the discrimination in the delivery of Facebook ad delivery. The second one corresponds to the unfairness against third-party products and sellers in the related item recommendations on the Amazon e-commerce platform.

4.2.1 Discrimination in Facebook ad delivery

In their classic work reported in ACM CSCW 2019 [1], the authors investigated the ad delivery aspect of the advertising platform of Facebook. Any ad on Facebook is always tied to a *page* and multiple pages can be utilized by an advertising agency to run their ads. The key components of this platform are as follows.

- *Ad contents*: the content of the ad comprises a headline and text accompanying the ad along with an image or video to be shown to the user. Sometimes the ad may also contain a destination URL.
- *Audience selection*: audience selection (aka ad targeting) can be done on Facebook in different ways. First, the advertisers can target audiences based on their demographic properties like age, gender, location, or profile characteristics. Second, exact users can also be targeted based on their personally identifiable information such as name, date of birth, etc. Third, they can target users similar (identified by Facebook) to those whom they had already targeted earlier. These schemes can also be used in combination with one another.
- *Bidding on Facebook*: on Facebook, advertisers can bid for different objectives optimizing for either the number of views/impressions, for click-through, or for the sales generated by the clicks. Further, the bid can either have a specified start and end time or it could have a daily budget cap.
- *Ad auction*: as and when slots are available, Facebook runs an auction to select an active ad for that user. The auction is decided based on the value of the bid, the engagement of the users with the ad so far, and the relevance of the ad.
- *Interface*: Facebook provides an interactive interface to advertisers for easy launching and monitoring of the ad in their delivery stage. In addition, it also provides a number of useful statistics to the advertisers which can further help them to fine-tune their ad targeting scheme.

4.2.1.1 Data curation

Custom audience preparation. As a first step for their experiments, the authors had to select a set of audiences. The authors performed most of their experiments on custom-generated audiences as follows. They randomly generated 20 lists of 1 000 000 distinct, valid North American phone numbers of the form +1 XXX XXX XXXX based on known valid area codes. For each list, Facebook could match ∼220K users.

Ad campaign and data collection. Next, the authors ran the ad campaign and collected the delivery performance statistics every 2 min using the Facebook Marketing API. The authors sought the data broken as per the attributes of the study including age, gender, and location. Among all the fields returned by the API, the authors make use of reach which is the number of unique users to whom the ad was shown. All the results that they report are in terms of reach; for instance—the fraction of men in the audience would refer to the reach of men normalized by the reach of men plus the reach of women.

4.2.1.2 Effect of ad content on ad delivery
Experimental setup: as a first experiment the authors find out the impact of the different types of content—headline, text, image—in the dissemination of the ad. In order to do this, the authors created two ads—*bodybuilding* (stereotypical of men) and *cosmetics* (stereotypical of women). The two ads had identical bidding strategies and budgets and were run at the same time. For each content type for an ad, different custom audiences were targeted, i.e., different custom audiences for (i) 'base' type, (ii) 'text' type etc. Base type ad has only a link with no headline, no text, and a blank image. Text type contains base type plus the text and so on. The richest form of content is 'image' which has the body, the text, the headline, and a relevant image. Note that the authors *did not target users based on gender*; the only constraint was that the age of the user should be 18+ and from the US.

Key results: the authors observed huge differences in the ad delivery patterns although the bidding strategy for all the ads was the same and both the ads were targeted to the same set of audiences irrespective of their gender. The 'image' type bodybuilding ad was delivered to men over 75% of the time on average. On the other hand, the 'image' type cosmetics ad was delivered to women more than 90% of the time on average. The authors further studied the effect of the different types of content. For the 'base' type, the bodybuilding ad reached 48% men and the cosmetics ad reached 40% men. For the 'text' type ad these percentages were very similar to the 'base' type. The moment the images were added to the ad content, the differences in the delivery became extreme.

Reasons: as a reason for this observation, the authors hypothesized that Facebook ran its own automatic text and image classification algorithms on the ad content to identify the relevance score agnostic to the ad's initial parameters and performance. To validate this hypothesis the authors considered the *alpha channel* present in most image formats and added 98% opacity. This would then not be discernible to the human eyes but an algorithm will still produce the same relevance score as for the unperturbed image. The authors chose five images that would be stereotypically of interest to men and five images that would be stereotypically of interest to women. For each of these, they also added 98% opacity and created another ten images. As a control, the authors also used five completely white images. For a total of these 25 images, the authors ran the ads with exactly the same bidding strategy, constant base, text, headline, and the audience is targeted such that each of them is potentially exposed to three ads (expecting them to see one masculine, one feminine, and one blank image). The authors observed that even when the images were *invisible* (to the human eye), the skew in ad delivery existed. The *invisible* male images reached significantly more male audiences on average. Similarly, the *invisible* female images reached significantly more female audiences on average. Since these images were invisible to the human eye, Facebook should have been actually using an algorithm to identify the relevance score of an ad for a user thus corroborating the hypothesis of the authors.

4.2.1.3 Racial skew in ad delivery
In their next set of experiments the authors studied the racial skew in ad delivery patterns.

Entertainment ads: here the authors constructed ads that led to lists of best albums —general top 30, top country music (stereotypically popular among white users), and top hip-hop albums (stereotypically popular among black users). Surprisingly, the authors found that the Facebook ad delivery conforms to the stereotypical distribution despite all ads having the same bidding strategy and being targeted in the same manner. In particular, 80% white users received country music ads while only 13% of them received hip-hop ads. This experiment showed that the relevance scores computed by the Facebook ad delivery algorithm somehow factors racial information into it.

Job ads: subsequently the authors went on to investigate if racial skews could be observed in more serious matters like job ads. For this they crafted ads for 11 job types—artificial intelligence developer, doctor, janitor, lawyer, lumberjack, nurse, preschool teacher, restaurant cashier, secretary, supermarket clerk, and taxi driver. Each ad was designed to land on a real-world job listing page. Further each of the ads creative image contained one of the five photos—photo of black female/male, or white female/male or an appropriate ad for the job type but with no photo of a person in it. All the ads used the same bidding strategy and targeted the same audience. After running the ads for just 24 h the authors found drastic differences in the ad delivery patterns with huge racial skews. The five ads pertaining to the lumberjack industry were delivered to 90% men and 70% white users in aggregate. On the other hand, the five ads pertaining to janitors were delivered to 65% women and 75% black users in aggregate. These results led the authors to conclude that ad delivery optimization by the delivery platform is in play and this supersedes the preferences set by the advertiser.

Housing ads: in this set of experiments the authors created a suite of housing ads by varying the type of property (purchase versus rental) and the implied cost (cheap versus luxury). Once again, despite the targeting parameters and the bidding strategy being the same for all ads, a huge racial skew was observed by the authors. While certain types of housing ads reached 72% of the black users, other types of ad reached as low as 51% of black users.

4.2.2 Unfairness against third-party products and sellers in Amazon

Information access (IA) systems are commonplace these days in order to mediate the smooth interaction between content producers and consumers. Such IASs have been deployed across various online platforms ranging from social media to e-commerce. While such IASs have been highly beneficial, they come with their own caveats. Most of them focus primarily on relevance and persornalization which often results in curtailment of equal opportunity or satisfaction, the emergence of a filter bubble, or the manifestation of information segregation etc. Multiple research studies have been performed both within platform organizations and by third parties to audit such IASs to understand the ramifications of the above listed caveats.

4.2.2.1 Audits for evaluating well-being of users
Equality of satisfaction among demographics: search engines are massively advertised to be available to all users regardless of their demographics. An internal audit of Microsoft's Bing search engine was conducted by researchers at Microsoft to understand the level of satisfaction of users across different demographics. Through their internal and external audit methods grounded on causal inference literature, they did not find any significant difference between user satisfaction across different genders. However, they found older users to be slightly more satisfied than younger users [2].

Political biases or lack thereof in search results: for politics related queries, search results of Twitter were found to induce bias toward the corresponding political affiliations of the query terms [3]. They further developed an easy to interpret quantification metric for political search bias by differentiating the input bias, output bias, and bias due to the ranking system. This method was further utilized (with slight variation) to quantify and demonstrate political bias in Google search results [4] where the authors found slight evidence of the 'filter bubble' hypothesis too.

More on 'filter bubble' effects on IA systems: 'Filter bubble' is defined as a phenomenon resulting in a self-reinforced pattern of narrowing down exposure [5]. This is essentially the manifestation of a rabbit hole which is formed due to consuming similar content over and over again and, thereby, giving signals to information access systems about the relevance of such content. Given that the IA systems are keyed to relevance, understanding the pattern that they further recommend content of a similar nature and pushing the consumer further down the rabbit hole. Such inadvertent effects are usually due to lack of serendipity and/or diversity in the recommended set of content. An interesting study on YouTube found that YouTube's recommendation algorithms frequently suggest alt-lite and intellectual dark web content. From these two communities, it is possible to find alt-right content from recommended channels as well thus depicting a picture of a radicalization pathway on YouTube [6].

While the discourse on algorithmic audit has traditionally been centered around the well-being of the users (especially consumers and rarely producers), recently a lot of attention is being put toward the role of the platform organizations themselves. Researchers have been looking into these aspects, particularly, in the context of e-commerce platforms such as Amazon.

4.2.2.2 Case study on Amazon e-commerce platform
In a recent study presented in ACM FAccT [7] on Amazon.in, the authors identified a number of special relationships in the marketplace due to several vertical integrations of Amazon.

- Amazon produces products under its own private label brands—**Private Label products** (PL). Examples include AmazonBasics, Amazon Brand Solimo, etc.
- They offer fulfillment services to other sellers under the **Fulfilled by Amazon sellers** (FBA) scheme. This makes many of the competitors of Amazon their customers.
- In India, Amazon has launched joint ventures with certain other retailers to form **Special Merchants** (SM) so as to sell their own products. This alliance is owing to the Indian law prohibiting Amazon to directly sell their products on their platform [8, 9].

Sponsored recommendation as a trojan horse to nudge toward PL: emergence of private label product has introduced a massive monetary incentive for platforms (like Amazon) to promote their own products at the cost of promotion of several other third-party products. This has been to the extent that Amazon was asked explicit questions about their treatment of private labels during the antitrust subcommittee hearing in the US [10, 11]. Even more worrying, Amazon has been gradually replacing organic search results by sponsored advertisements on its product pages [7]. By creating a related item network (RIN), the authors in this study found that Amazon sponsored recommendations were being used to systematically nudge customers more and more toward PL products in the examined categories.

A RIN is a directed network where each node is a product/item and an edge $p_i \rightarrow p_j$ indicates that the product p_j is recommended on the product page of p_i. To distinguish between the RINs, the authors referred to the RIN constructed from sponsored recommendations as sponsored RIN (S) and that from the organic recommendations as organic RIN (O). In each of these networks, the products can have one of two labels—private label (PL) or a third-party (3P). Based on some simple analysis of these networks, the authors observed various drastic differences in the way PL and 3P products are promoted in the sponsored recommendations. The first among these was the average in-degree of a node p_j, i.e., the number of product pages in which p_j features a recommendation. In one of the two categories (i.e., battery) the authors investigated, PL products appeared as organic recommendations on the product page of 46 other items on average. In contrast, PL products appeared as sponsored recommendations on the product page of as high as 520 other items on average. Similarly, for the other category, PL products featured as organic (sponsored) recommendations on the product page of 45 (164) other items on average. On the other hand, for the 3P battery category, a product appeared as a sponsored recommendation on the product page of only 9 other items on average. This means that in sponsored recommendations, it is 58 times (i.e., 520/9) more likely for a PL product to be featured compared to a 3P product. Similar skew was observed for the backpack category of products also. The authors also performed *k*-core decomposition [12, 13] of the two RINs and observed the following. In the sponsored RIN, if the cores are divided into four quartiles, 94% of PLs were in the top two quartiles for the battery category; for the backpack category, this was 67%. In the organic RIN, only 65% (57%) PLs were found in the first two quartiles for the battery (backpack) category. Finally, the authors also performed random walk on both the RINs and observed the following. The authors term the stationary probability of the random walk as the *exposure* that a product gets. They observed that in the sponsored RIN, Amazon PLs see a 160% rise in exposure compared to the organic RIN. They also found that over 75% of the 3P products were under-exposed in the sponsored RIN.

Disparity in buy box outcomes toward SMs: another important aspect of e-commerce platform (esp. Amazon) is the 'buy box' of Amazon. It is a tiny box shown on the product page of any Amazon product which includes two buttons that read 'Add to cart' and 'Buy now'. When multiple sellers compete to sell a specific product, a proprietary algorithm decides which seller will win the auction for the buy box; and thus get featured on the product page. Increasingly, Amazon's buy box algorithm has been under scrutiny by policymakers in the US [10] and EU [14] because it can be used to avoid competition and nudge users toward sellers having special relationships with the marketplace. Upon an investigation of over 35 K such buy box competitions of different products (at different time stamps), we observed that there is a significant skew in the percentage of buy box wins when Amazon SMs compete. In fact, one can say whenever they compete inevitably they win the buy box.

4.3 Content moderation

Social media has immensely facilitated the formation of distant connections, new groups and collaborations, and the instant sharing of messages/opinions. However, this has resulted in many bad actors spreading harmful, inappropriate, and fake content. Such content can mentally affect the community as well as having the potential to sway public opinion. In many cases the online content has been found to have many offline consequences. For instance, a piece of fake news reporting the injury of Barack Obama in an explosion wiped out $130 billion in stock value [15]. Likewise, numerous fake stories became viral during the 2016 US presidential election. Strong connection between harmful content and hate crime has been often reported, e.g., the Rohingya genocide, the Sri Lankan mob violence[1], and the Pittsburg shooting[2] to name a few. It is therefore imperative to take necessary steps to reduce the spread of such content over social media. In fact, recently, a UN report [16] was prepared to strategically mitigate hate speech in the online world.

Content moderation refers to the screening of user-generated content before being published on online platforms. The content is screened to ensure that it is not

---

[1] https://www.hrw.org/news/2019/07/03/sri-lanka-muslims-face-threats-attacks
[2] https://en.wikipedia.org/wiki/Pittsburgh_synagogue_shooting

hateful, inappropriate, or fake in nature. In the initial days such screening was mostly manual. However, with the prolific increase in the volume of content posted daily, it has become impossible to manually sieve through all of these. Recently, therefore machine learning algorithms are being extensively used to filter all forms of inappropriate and illegal content [17] and then present only the 'difficult to identify' cases to the moderators for manual inspection. This extra layer of automatic moderation drastically reduces the effort as well as the mental toll experienced by the moderators in handling such content.

In the following sections, we shall introduce the problem of fake news, hate speech, and media bias along with the automatic ways to handle these problems that are plaguing social media and potentially resulting in real-world violence.

4.3.1 Fake news

Fake news is deliberate and verifiable false news published by a news outlet. The intention of fake news is usually bad. The literature here is crammed with various forms of news types varying in intention and authenticity. In table 4.1 we present a classification of the most common such news types. There are primarily two major theories that guide the investigation of fake news. These are noted and discussed as follows.

4.3.1.1 Style based theories

This theory pertains to studying fake news based on the style in which it is written. It builds on the hypothesis that fake news is written in a linguistic style that is different from authentic news. There are three prominent types of theory in place here which are summarized below.

- *Undeutsch hypothesis*: a statement that is based on facts that differ in its content and quality from a statement borne out of fantasy.
- *Reality monitoring*: true events trigger higher levels of sensory-perceptual information compared to false ones.
- *Four-factor theory*: truth differs from lies in the way it is expressed in terms of arousal, behavior control, emotion, and thinking.

Table 4.1. Classification of news types.

Type	Authentic	Intention	News
Fake news	False	Bad	Yes
False news	False	Unknown	Yes
Satire news	Unknown	Not bad	Yes
Disinformation	False	Bad	Unknown
Misinformation	False	Unknown	Unknown
Rumor	Unknown	Unknown	Unknown

4.3.1.2 Propagation based theories
This theory suggests the study of fake news from the perspective of how it spreads on social media. Once again there are some popular prototypes within this theory. These are summarized below.

- *Backfire effect*: evidence countering the inherent beliefs of individuals can result in them rejecting the evidence even more strongly.
- *Conservatism bias*: individuals might be conservative to the point that they only insufficiently revise their beliefs when presented with new evidence.
- *Semmelweis reflex*: often, individuals tend to reject new evidence that contradicts what they believe is normative and traditional.

The outcome of both these theories is that it is almost impossible to rectify the views and beliefs of individuals once fake news has gained their trust. In other words, fake news is inherently incorrect and hard to correct. Thus the only solution lies in detecting and containing fake news as early as possible, i.e., before it is able to build trust among individuals. We describe below in brief the different computational approaches to identify fake news at an early stage. As there are numerous studies in this space, we shall briefly outline some of the main methods referring the reader to relevant articles for further information.

4.3.1.3 Fake news detection
Some of the most important techniques available for fake news detection are as follows.

- Knowledge based fake news detection.
- Style based fake news detection.
- Propagation based fake news detection.

We now discuss each of these approaches briefly.

Knowledge based fake news detection: in this approach, known facts, i.e., true knowledge, are compared with incoming news content to ascertain the authenticity of this content. This is also known as *fact checking*. There are two primary types of fact checking—(a) manual and (b) automatic. Each of these have their own pros and cons. Manual fact checking is usually costly and requires expert intervention. Automatic fact checking is scalable and is a better choice for application development. Automatic fact checking usually has three challenges—(i) representation of knowledge, (ii) extraction of known facts (aka ground truth), and (iii) comparison of known facts with the to-be-verified content.

Usually, knowledge is represented as a set of SPO (subject, predicate, object) triple extracted from the information available as input. For instance, the statement 'the tree has the color green' comprises the subject 'the tree', the predicate 'has the color', and the object 'green'. Such SPO triples are typically represented as a graph where the subject and the object form the nodes and the predicate defines the edge. Such a graph is called a *knowledge graph*. A hypothetical example of a knowledge graph is shown in figure 4.1.

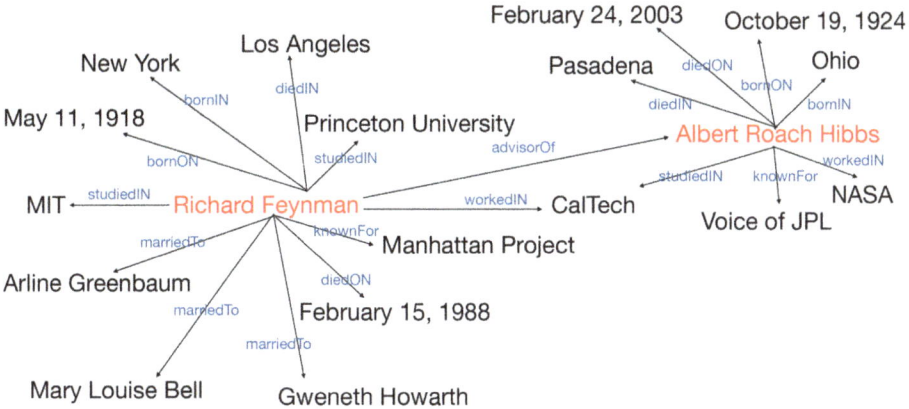

Figure 4.1. Hypothetical example of a knowledge graph.

The first step to fact checking is to extract such SPO triples and represent them in a knowledge graph. Such graphs can be constructed by gathering web content present in some relational form. The task of gathering this content can be done in different ways, e.g., by experts or volunteers manually, by automatic extraction of semi-structured documents like Wikipedia infobox, or by automatic extraction of completely unstructured text using state-of-the-art NLP techniques. The automatic extraction passes through a series of steps as follows.

- *Entity resolution*—detects mentions which correspond to the same real-world entity, e.g., (Abdul Kalam, profession, PresidentofIndia) and (APJ Abdul Kalam, profession, PresidentofIndia) constitute a redundant pair and need to be mapped to a single entity. Distance [18–20] or dependence [21–23] based techniques are currently used to tackle this problem.
- *Time recording*—remove outdated knowledge. These cases are tackled by inspecting the beginning and the end time of a fact, e.g., (UK, joinIn, EU) can be identified as outdated by associating the start and end dates to this fact.
- *Knowledge fusion and credibility evaluation*—identify and resolve conflict resolution. For instance (Abdul Kalam, bornIn, Tamil Nadu) and (Abdul Kalam, bornIn, Kerala) are two conflicting pieces of information and need to be resolved. Such conflicts are resolved by resorting to acquiring information from additional websites with high credibility.
- *Knowledge inference* based on link prediction techniques. Popular methods include latent feature models [24, 25], graph feature models [26], and CRFs [27, 28].

The second step is to compare the knowledge present in the knowledge graph with the information extracted from a to-be-verified news item. Let the facts from the knowledge graph be represented as $\langle s, p, o \rangle_{KG}$ and the extracted information from the news article be $\langle s, p, o \rangle_{news}$. A straightforward scheme to make this comparison is shown in algorithm 1. In the algorithm, CWA refers to the closed-world assumption (i.e., the truth value of a statement is true only if it is known to be true), OWA refers to the open-world assumption (i.e., the truth value of a statement may be true irrespective of whether or not it is known to be true) and LCWA refers to the local closed-world assumption which requires that if a knowledge graph contains one or more object values for a given subject and predicate, then it should contain all the possible object values. Now even if one such triple is present in the knowledge graph then the to-be-verified triple $\langle s, p, o \rangle_{news}$ cannot be true.

Style based fake news detection: style based fake news detection employs hand-crafted machine learning features that are able to distinguish fake news from true news. The motivation comes from the *Undeutsch hypothesis* [29] which states that the writing style for fake news is very different from true news. Usually, the writing style is captured through a set of linguistic properties that could beat the *lexical*, the *syntactic*, the *semantic* and the *discourse* level. The problem is formulated as follows. Let us assume that the to-be-verified news article can be represented by a set of linguistic features **F**. Based on a collection of training data, the objective is to learn a function $\mathcal{F}$ such that $\mathcal{F}: \mathbf{F} \longrightarrow \hat{y}$ where $\hat{y} = \{0, 1\}$ is the predicted label with 1 indicating fake news and 0 indicating true news. The function $\mathcal{F}$ is learned by optimizing a loss on the training dataset of the form $T = \{(F^{(i)}, y^{(i)}): F^{(i)} \in \mathbb{R}^d, y^{(i)} \in \{0, 1\}, i \in \mathbb{N}_+\}$. In T, each news article is represented as a d-dimensional feature vector $F^{(i)}$ associated with a known label $y^{(i)}$. The features are constructed

Algorithm 1: Algorithm to verify an information.

if $< s, p, o >_{news} \in < s, p, o >_{KG}$ **then**
$\quad < s, p, o >_{news}$ is true;
else
$\quad$ **if** *assumption=CWA* **then**
$\quad\quad < s, p, o >_{news}$ is false;
$\quad$ **end**
$\quad$ **if** *assumption=OWA* **then**
$\quad\quad < s, p, o >_{news}$ is true or false;
$\quad$ **end**
$\quad$ **if** *assumption=LCWA* **then**
$\quad\quad$ **if** $|< s, p, * >_{KG} > 0|$ **then**
$\quad\quad\quad < s, p, o >_{news}$ is false;
$\quad\quad$ **else**
$\quad\quad\quad < s, p, o >_{news}$ is true or false;
$\quad\quad$ **end**
$\quad$ **end**
end

from the lexical, syntactic, semantic, and discourse properties of the text. We describe these features briefly below.

- *Lexical features*: most of the lexical features include bag-of-words statistics like *n*-gram frequency and TF-IDF.
- *Syntactic features*: syntactic features include two subtypes—the shallow syntactic features and deep syntactic features. Shallow features include the frequency of parts-of-speech tags and named entity tags. Deep features include the frequency of the rewrite rules extracted from parse trees built using PCFG (or equivalent) grammar.
- *Semantic features*: at the semantic level various psycho-linguistic features like sentiments present in the news article etc are measured. In [30], the authors separately extracted semantic features from the title and the body of the news article. Specifically, the title was tested to identify if it was clickbait which is often true for fake news. This was done by a collection of different techniques including (i) comparison with a dictionary of clickbait-like titles, (ii) readability analysis (i.e., readability scores of clickbait titles are typically low), (iii) sentiment analysis (i.e., clickbait titles have an expression of extreme sentiments in them), (iv) testing the presence of special punctuation marks (e.g., '...', '?', '!') in the title, (v) quality analysis to find the similarity of the summary of the body of the news with the title (should be high for true news), and (vi) informality analysis (e.g., presence of swear words—'shit', netspeak—'lol', assents—'OK', non-fluencies—'hmm' etc). Similarly, from the body of the news article, different features were collected including (i) informality (same as for title), (ii) diversity, i.e., the number of unique words, content words, nouns, verbs, adjectives, and adverbs being used in news body, (iii) subjectivity that computes the number of biased words (taken from [31]) in the article, and the number of fact denoting words (e.g., 'observe', 'announce' etc) in the article, (iv) sentiment content, i.e., the proportion of positive and negative words as well as the overall polarity, (v) quality which computes the number of characters, words, sentences, and paragraphs as well as the average characters per word, words per sentence, sentences per paragraph, and (vi) specificity as indicated by certain LIWC features.
- *Discourse features*: at the discourse level the authors computed the frequencies of rhetorical relationships obtained as rewrite rules of the RST parser [32].

Propagation based fake news detection: propagation based fake news detection exploits the information on how fake news spreads in the social network, who spreads them, and how people spreading fake news are connected among themselves. One of the most heavily studied properties includes the *cascade phenomenon*, i.e., how news items/tweets are shared and re-shared in the social network. The authors in [33] used the news cascade to identify user roles (e.g., opinion leaders versus normal users) and observed that while true news is initiated by an opinion leader and then shared and re-shared by a lot of normal users, fake news is typically initiated by a normal user and then somehow re-shared by one or two opinion leaders which make them go viral in the network. Each share/re-share is encoded using the stance toward the news item (approval versus doubt) and the sentiment about the news item expressed by the users. The authors used a random walk based graph kernel to measure the similarity among the news cascades. Their assumption was that fake news can be distinguished from true news since the fake news cascades would be similar among themselves and different from the true news. In another study [34], the authors modeled news cascades as multivariate time series and employed RNNs as well as CNNs to detect fake news. Other forms of network structures that indirectly indicate the propagation of news in social networks have also been used in the literature. For instance, in [35] the authors use relationships among the news articles, the publishers, the users who share and re-share the news, and their accompanying posts to construct graphs and use a PageRank variant to identify fake news. Along similar lines, tensor decomposition [36, 37] as well as deep neural models like RNNs [38, 39] have been proposed to detect fake news. Another form of network that has been used is the stance graph built based on the for-against stance information [40].

4.3.1.4 Standard datasets for fake news detection
In this section we shall briefly outline some of the most important annotated datasets available for evaluating fake news detection models. Note that this list is not exhaustive and there are newer datasets being built and released quite frequently.

- *PolitiFact* [41, 42]: this data is based on politifact.com which hosts fact-checked statements appearing in news articles related to American politics. The statements can have labels like 'true', 'mostly true', 'false', 'mostly false' etc.
- *Snopes* [43]: the data is obtained from snopes.com, which contains political and other social media content. The news items and associated video are fact-checked and placed in classes like 'true', 'mostly true', 'false', 'mostly false', 'unproven', 'outdated', 'scam' etc.
- *FactCheck* [44]: the data again is about American political news hosted at https://www.factcheck.org/. Different items like TV ads, speeches, debates, news, and interviews are fact-checked and organized into three classes - 'true', 'no evidence', and 'false'.
- *LIAR* [45]: this is a dataset available in the public domain curated from politifact.com. It has 12.8 K manually labeled short statements collected in different contexts for over a decade. Along with the labels (exactly as in the PolitiFact dataset), each data point also has a justification for the label.
- *COVID-19 misinformation dataset* [46]: the dataset comprises false claims and statements that seek to challenge or refute them. The authors train a classifier to build a new dataset comprising 155 468 COVID-19-related tweets, with 33 237 false claims and 33 413 refuting arguments.
- *Fakeddit* [47]: this is a novel multimodal dataset curated from Reddit that contains 1 M entries from multiple categories of fake news. The dataset has different granularity of labels—2-ways, 3-ways, and 6-ways. The six fine-grained classes include 'true', 'satire', 'misleading content', 'imposter content', 'false connection', and 'manipulated content'.

- *Weibo21* [48]: this dataset is curated from Sina Weibo containing 4488 items of fake news and 4640 real news from nine different domains. The different domains considered by the authors were 'science', 'military', 'education', 'disasters', 'politics', 'health', 'finance', 'entertainment', and 'society'. The dataset has a variety of content including text, pictures, time stamps, comments, and judgment information.
- *MuMiN* [49]: this is a graph dataset built from 21 M social media posts from Twitter including tweets, replies, users, linked articles images, and hashtags. The data comes from 26 000 distinct Twitter threads connected to 13 000 fact-checked articles across different topics and different languages.
- *TruthOrFiction* [50]: this data is sourced from the fact checking website https://www.truthorfiction.com/. The content includes Internet rumors, urban legends, and other deceptive stories or photographs. The content spans multiple topics like politics, religion, nature, food, medical, etc and the labels are primarily 'truth' or 'fiction'.

4.3.1.5 State-of-the-art results in fake news detection

The results in the fake news detection literature primarily vary based on the dataset, the number of classes, and the model complexity. The authors in [30], report a macro F1-score of ~0.89 on PolitiFact dataset and ~0.88 on BuzzFeed dataet. They also observed that the lexicon and the deep syntax level features are the most discriminative. The authors in [33] collected data about known false rumors from the Sina community management center and evaluated the performance of their random walk kernel based model on this dataset. They reported an F1-score of ~0.91 in both classes. Using different variants of CNNs the authors in [45] achieved a maximum accuracy of ~0.27. The authors in [46] showed that the F1-score for COVID-19-related misinformation in two different topics were ~0.78 and ~0.69 respectively. On the Weibo21 dataset, the authors [48] reported an F1-score of ~0.91 using a multi-domain model built up on the BERT architecture.

4.3.2 Hate speech

One of the most notorious problems in the online world is the burgeoning growth of harmful content targeting marginalized individuals and communities. The problem is so critical that in 2019, the United Nations proposed the Strategy and Plan of Action on Hate Speech observing the worldwide 'disturbing groundswell' of racism, xenophobia, and religious intolerance including anti-Semitism, Islamophobia, and persecution of Christians. An executive committee was set up in view of this glaring problem; the committee came up with the following recommendations—(i) a worldwide initiative for monitoring and analyzing hate speech, (ii) addressing the root causes and drivers of hate speech, (iii) using technology to detect hate speech, and (iv) educating people to address and counter hate speech. Researchers, practitioners, and policymakers across the globe have started making collective efforts to combat this online (and offline) evil. The research in this area can be organized into three distinct types as follows.

- Analysis of the spread of hate speech.
- Automatic detection of hate speech and its variants.
- Mitigation of the effects of the detected hate speech.

In the following, we shall discuss each of these in detail. However, we shall first outline the operational definition of hate speech which is needed for the computational treatment of the problem.

4.3.2.1 Definition of hate speech
Hate speech is defined as direct and serious attacks on any protected category of people based on their race, ethnicity, national origin, religion, sex, gender, sexual orientation, disability, or disease. There are broadly two categories of hate speech as follows.

- *Directed hate*: hate language toward a specific individual or entity. Example '@usr4 your a f*cking queer f*gg*t b*tch'.
- *Generalized hate*: hate language toward a general group of individuals who share a common protected characteristic, e.g., ethnicity or sexual orientation. Example: '—was born a racist and—will die a racist!—will not rest until every worthless n*gger is rounded up and hung, n*ggers are the scum of the Earth!! wPww WHITE America'.

The targets of hate speech depend on the platform, demography, and language and culture [51]. There has been focused research on characterizing such diverse types. Examples include racism against blacks in Twitter [52], misogyny across the manosphere in Reddit [53], Sinophobic behavior with respect to the COVID-19 pandemic [54]. Further, hate speech often becomes part of different communities; examples include the genetic testing conversations [55] and the QAnon conversations [56]. In the following, we shall attempt to present a comprehensive footprint of the major research activities in hate speech.

4.3.2.2 Spread of hate speech
In view of the increasing threat manifesting from online hate speech, traditional social media platforms like Twitter, Facebook, YouTube, Instagram, etc, have resorted to heavy moderation. As a result of this, alternative platforms like Gab[3], 4Chan[4], BitChute[5] etc have come into existence. These websites have reportedly welcomed alt-right and extremist users many of whom were banned from the traditional social media platforms[6]. These alternative platforms have lax moderation and portray themselves as 'champions of free speech'. Here we shall discuss

---
[3] https://gab.com/
[4] https://www.4chan.org/
[5] https://www.bitchute.com/
[6] https://www.adl.org/blog/when-twitter-bans-extremists-gab-puts-out-the-welcome-mat

some of the important works that pointed out the prevalence and spread of significant harmful content on these platforms.

In [57], the authors collected 22 M posts from 336 k users crawled from gab.com, between August 2016 and January 2018. A simple count of the frequency of hashtags showed that items like #Alt-right, #BanIslam are very popular. In [58], the authors compared the cascade properties of the content spread by hateful and non-hateful users. They collected 21 M posts from 340 k users, between October 2016 and June 2018 from gab.com. They started with a small set of hand-labeled hateful users and then used a diffusion model to infer the hatefulness of the neighbors of these users. They first built a repost network as shown in figure 4.2(a) where each node is a user and a directed edge from user u_i to user u_j with weight w_{ij} indicates that u_i has reposted w_{ij} messages of u_j. A weighted self-loop indicates how many messages the user has directly posted. This repost network is then converted to a belief network as shown in figure 4.2(b). For instance, since user B in the figure posts 5 messages and reposts 15 messages from user A, the belief of user B on themselves is $\frac{5}{(5+15)} = 0.25$ and on user A is $\frac{15}{(5+15)} = 0.75$. Note that the direction of the edges has naturally been reversed in this network which now represents the belief of a user on another. The authors then run the DeGroot model [59] of information diffusion on this belief network starting from those nodes that have been manually labeled as hateful (see figure 1.3(c)) and are assigned a hatefulness score of 1. In this initial step, the hatefulness of all other nodes is set to 0. In each step of the diffusion process, the hatefulness scores of all the nodes are recalculated. For instance, in figure 1.3(c), the hatefulness scores of node A (manually labeled as hateful), B and C in the initial step are $b_A^{(0)} = 1.0$, $b_B^{(0)} = 0.0$ and $b_C^{(0)} = 0.0$ respectively. In the next step, the score of node C would be $b_C^{(1)} = 0.4 b_C^{(0)} + 0.5 b_A^{(0)} + 0.1 b_B^{(0)} = 0.4*0.0 + 0.5*1.0 + 0.1*0.0 = 0.5$. At convergence, each node has a score indicating the extent of the hatefulness of the corresponding user. The authors classify nodes having a score [0.75, 1] as hateful (KH) users (2290 in number) and those having a score [0, 0.25] as non-hateful (NH) users (58 803 in number). Once all the nodes are labeled in this

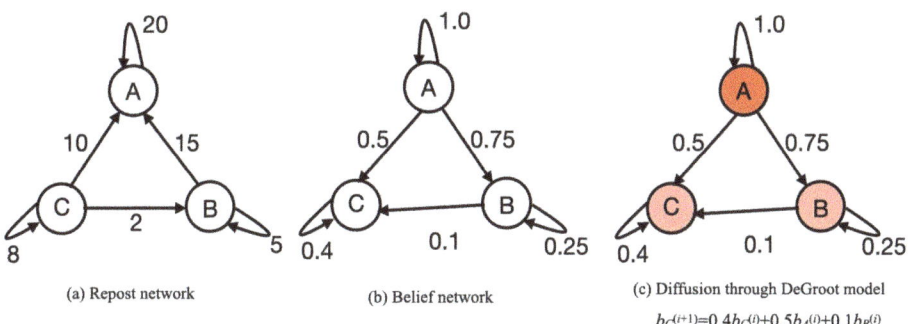

Figure 4.2. Identification of hateful users based on the DeGroot model of diffusion. Node A marked in dark red indicates the manually labeled hateful user. The nodes B and C marked in light red indicate the extent of hatefulness inferred by the diffusion model at convergence.

way the authors verified the effectiveness of their algorithm through human judgments and found that the classification is highly accurate.

They then studied information cascades that are formed through these labeled nodes. In order to build the cascades they constructed a different variant of a repost network where the nodes still represent users but the edges represent followership links (see figure 4.3(a)). The numbers on the nodes represent the time of repost by the node. The idea was to identify the node that unambiguously influences another node. However, this, as the authors noted, could be tricky. For instance in figure 4.3(a), it is unclear whether the repost by C at time 250 is influenced by A or B (both of whom are in receipt of the message before C and both of whom C follows). The authors resorted to the popular LRIF (least recent influencer) heuristic [60] to resolve this ambiguity thus assuming C to be influenced by the node which had received the message least recently (A in this case)[7]. Application of this heuristic results in a DAG (directed acyclic graphs) which is exactly what is needed to study the cascade properties.

The authors collected all the DAGs obtained from every individual message that was initiated by a KH user and compared their aggregate properties with those collected for the NH users. Following prior literature [61], the standard cascade properties that they studied were—(i) *size*, i.e., number of unique users, (ii) *depth*, i.e., length of the largest path in the cascade, (iii) *average depth*, i.e., the average depth of each path from the root node, (iv) *breadth*, i.e., the maximum width of the cascade, and (v) *structural virality* which indicates viral nature of the post. The aggregate behavior of the KH versus NH cascades is shown in table 4.2. The results are shown for all posts as well as posts with multimedia content and posts in different topics[8]. For all three categories, the authors observed that the content initiated by the hateful users has a larger audience, spread wider and deeper, and is

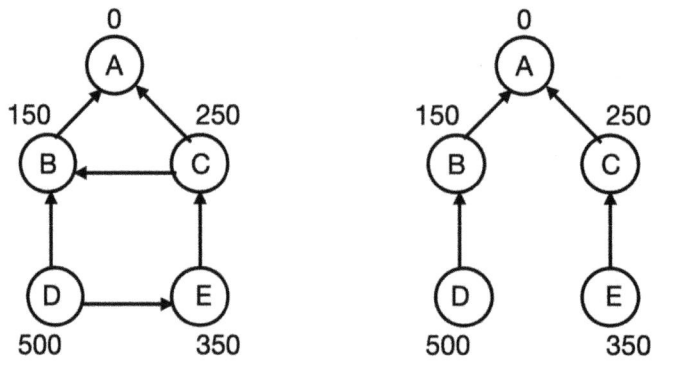

(a) Flow of message across the followership network (b) DAG obtained based on LRIF model

Figure 4.3. Cascade of information over the followership network.

---

[7] The other alternative is the MRIF (most recent influencer) heuristic that would designate B as the influencer of C.
[8] KT: a higher proportion of hateful content; NT: those having a lower proportion of hateful content.

Table 4.2. The different cascade properties for the posts initiated by KH and NH users.

Feature	Posts		Attachments		Topics	
	KH	NH	KH	NH	KT	NT
Size	**1.28**	1.21	**1.34**	1.23	**1.68**	1.51
Depth	**0.13**	0.09	**0.16**	0.11	**0.30**	0.24
Breadth	**1.13**	1.10	**1.15**	1.11	**1.30**	1.24
AD	**0.11**	0.08	**0.14**	0.10	**0.26**	0.22
SV	**0.13**	0.09	**0.16**	0.11	**0.31**	0.25

more viral. The differences in the distribution of all these properties are statistically significant.

In a follow up work [62] the authors studied the temporal evolution of hateful users/content in the Gab network. They found that overall the amount of hate content is increasing over time. Further, hateful users are found to occupy strategic positions in the inner core of the network more and more over time. Last, but not least, the overall linguistic style of the normal Gab user become more aligned with those of the hateful users as time progressed.

The authors in [63], analyzed the 4Chan platform using a dataset crawled from the website's/pol/thread between June 29, 2016, to November 1, 2019. They used the Perspective API to identify the extent of hatefulness in the content posted. The authors found that around 37% of the data had some element of toxicity while 27% was severely toxic. They also found that the platform supported various activist movements and alt-right political ideologies.

There have been some works on moderated platforms also. One of the most popular among these was [64] where the authors investigated the properties of hateful users on Twitter. The data was collected using a set of hateful keywords. The main observations of the authors were that the hateful users are *power users* (i.e., post more and favorite more) and a median hate user is more central to the network. Finally, a recent work [65] has identified a more nuanced form of harmful content—*fear speech* [65]—that is used to incite fear against a minority community. The authors showed that such posts are abundant in WhatsApp groups, do not contain toxic elements, and use an argumentative structure (often fake) to gain traction and evade platform moderation. The authors found that such posts enjoy more reshares and a longer lifetime. A large number of users spread them and a large number of groups are affected. More interestingly, certain emojis are used in groups along with a post to make the 'fear-effect' far-reaching.

4.3.2.3 Detection of hate speech
We will organize this section into three parts—the datasets available, the detection models, and the biases in these models.

Datasets: the datasets can be classified based on taxonomies, sources, and languages. As regards, taxonomy one can have a binary classification (hate/not hate, targeting a group or not) [64, 66–69], specific binary classification (misogyny/not, racism/not) [70–74], multiclass/multilabel datasets [75–77], etc. As regards sources, traditionally Twitter has been one of the primary ones [78–80]. Other traditional platforms include YouTube [67, 81, 82], Facebook [83, 84] etc. Of late, there have been many other (alt-right) sources like Gab [62, 70], Parler [85], Voat [86], 4Chan [87], BitChute [88] etc. As regards language, English has been the most predominant one [70, 76–78, 89, 90]. Recently, initiatives have been taken to build datasets in other languages including Arabic [91, 92], Italian [93, 94], Spanish [95, 96], and Indonesian [97, 98] each of which has more than three datasets. There have also been many efforts in building datasets in Indic languages like Hindi [99], Bengali [100, 101], Malayalam [102, 103], Kannada [104], Tamil [103, 105], Telugu [106], etc. Two new Korean datasets [107, 108] have also been released recently. The abusive portion in these datasets ranges roughly between 25% and 50%.

Detection models: the earliest models were based on feature engineering. Some of the standard features [109, 110] used were TF-IDF vectors, parts-of-speech tags, linguistic features like sentiment lexicons, frequency counts of URLs, username, and readability scores of the posts. These features were fed into standard off-the-shelf machine learning models like SVM, logistic regression, random forests, etc to predict whether the input text was hateful or not.

Subsequently, text embeddings were used to improve the performance of the above models. In particular, word embeddings [111], as well as sentence embeddings [112] constructed from tweets, have been used to detect hate speech. In parallel, newer models like XGboost [112], LSTM/GRU [113], and CNN-GRU [114] have been introduced that are able to take advantage of such embedding techniques.

With a surge in transformer architectures, many new models have come into existence that attempt to accurately detect hate speech. Most of these utilize the BERT framework and the authors in [115] performed a detailed analysis of the different variants of this framework. The main results of this comparison are noted in table 4.3. More recently the BERT model has been completely retrained from scratch using banned subreddit data and this new model has become popular by the name of HateBERT [116]. As shown in table 4.4 the HateBERT model outperforms the BERT base model for various benchmark hate speech datasets.

There have also been initiatives to develop multilingual hate speech detection models in the literature. One of the most popular among these was proposed in [115]. The authors presented a recipe for the type of model that is most appropriate for a particular data setting. In table 4.5 we reproduce this recipe which can be interpreted as follows—(a) monolingual: in this setting, data from the same language is used for training, validation, and testing; (b) multilingual: in this setting data from all but one language is used for training, while validation and test set are drawn from the remaining (i.e., target) language, (c) LASER+LR: for the input post, the authors first constructed a LASER embedding [120] which are obtained by applying max-pooling operation over the output of a BiLSTM encoder. These embeddings are then passed as input to a logistic regression (LR) model for the final prediction; (d) translation+BERT: the input post is translated to English using Google translator which is then provided as input to the BERT model for the final prediction;

Table 4.3. Comparison of BERT based models for the task of hate speech detection. The datasets used are the ones released in Waseem and Hovy [117] and Davidson et al [118]. The best performance is highlighted in bold font.

Method	Datasets	Precision (%)	Recall (%)	F1-score (%)
Waseem and Hovy [117]	Waseem	72.87	77.75	73.89
Davidson et al [118]	Davidson	91.00	90.00	90.00
Waseem et al [119]	Waseem	—-	—-	80.00
	Davidson	—-	—-	89.00
BERT base	Waseem	81.00	81.00	81.00
	Davidson	91.00	91.00	91.00
BERT base + Nonlinear Layers	Waseem	73.00	85.00	76.00
	Davidson	76.00	78.00	77.00
BERT base + LSTM	Waseem	87.00	86.00	86.00
	Davidson	91.00	92.00	92.00
BERT base + CNN	Waseem	**89.00**	**87.00**	**88.00**
	Davidson	**92.00**	**92.00**	**92.00**

Table 4.4. The comparison of the performance of BERT base model with the HateBERT model.

Dataset	Model	Macro F1	Pos. class—F1
OffensEval'19	BERT base	0.803 ± .006	0.715 ± 0.009
	HateBERT	0.809 ± 0.008	40.723 ± 0.012
AbusEval	BERT base	0.727 ± 0.008	0.552 ± 0.012
	HateBERT	0.765 ± 0.006	0.623 ± 0.010
HatEval	BERT base	0.480 ± 0.008	0.633 + 0.002
	HateBERT	0.516 ± 0.007	0.645 + 0.001

Table 4.5. A recipe of the model and data type combination that works best in the low and high resource settings. The performance in terms of macro F1-scores for each of these combinations are shown in the parentheses.

Language	Low resource (128 instances)	High resource (All instances)
Arabic	Monolingual, LASER+LR (0.75)	Multilingual, mBERT (0.84)
English	Monolingual, LASER+LR (0.56)	Multilingual, mBERT (0.74)
German	Monolingual, LASER+LR (0.59)	Translation+BERT (0.77)
Indonesian	Monolingual, LASER+LR (0.63)	Monolingual, mBERT (0.81)
Italian	Monolingual, LASER+LR (0.72)	Monolingual, mBERT (0.83)
Polish	Monolingual, LASER+LR (0.57)	Translation+BERT (0.72)
Portuguese	Monolingual, LASER+LR (0.62)	Monolingual, LASER+LR (0.69)
Spanish	Monolingual, LASER+LR (0.59)	Multilingual, mBERT (0.74)
French	Monolingual, LASER+LR (0.63)	Translation+BERT (0.66)

(e) mBERT: the input post is directly fed to the mBERT (multilingual BERT) model[9] for the final prediction. The table indicates for each language the authors experimented with, the best-performing combination of model and data type when the number of data points available is low/high. The detection performance in terms of the macro F1-scores for each of these combinations is also shown in the table.

Biases in the detection models: while a multitude of models has become popular, evaluations have often been notably spurious. For instance, in [121], the authors arrived at incorrect results due to oversampling before the train-test split was done. The correction of this error resulted in a drop of ~15% macro F1-score. Another work [122] reported spurious results due to feature extraction using the whole train and test split. The correction of this error resulted in a drop of ~20% macro F1-score. Further, there can be biases that arise from the data being used [118], the annotators involved in labeling the data [123], and the speaker/target community to whom the hateful speech is aimed [124]. This makes it necessary that the predictions made by the models are explainable. Typically the deep neural models function as a black box and it is difficult to interpret the predictions of the model (see figure 4.4 (left)). In order to circumvent this problem, two recent works [76, 125] proposed having *rationales* annotated in the data along the labels and targets. Rationales are spans of text in a post annotated to indicate the reason for an annotator to consider the post as hateful (see figure 4.4 (right)). The HateXplain [76] data consists of 20 k posts from Twitter and Gab that were annotated with labels, targets, and spans which represent the rationale in the posts. The ToxicSpans [125] consists of 10 k toxic posts from civil comments that were annotated with spans that represent the rationale in the posts.

The authors in [76] built a ground truth attention vector from the rationales and used it to supervise the attention of the [*CLS*] token in the final layer of the BERT model. They called this model BERT-HateXplain and compared its performance with standard attention-based baselines (the key results are reproduced here in

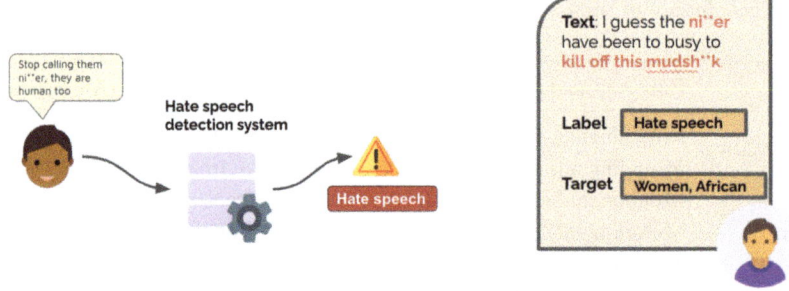

Figure 4.4. Explainable hate speech detection framework.

---

[9] https://github.com/google-research/bert/blob/master/multilingual.md

Table 4.6. Performance of BERT-HateXplain in comparison to other attention-based baselines.

Models	Accuracy	F1-Score	AUROC
CNN-GRU	0.627	0.606	0.793
BERT	0.690	0.674	0.843
BERT-HateXplain	**0.698**	**0.687**	**0.851**

Table 4.7. Bias metric based performance of BERT-HateXplain in comparison to other baselines.

Models	GMB-Sub	GMB-BPSN	GMB-BNSP
CNN-GRU	0.654	0.623	0.659
BERT	0.762	0.709	0.757
BERT-HateXplain	**0.807**	**0.745**	**0.763**

table 4.6). However, the more important point is that BERT-HateXplain is less biased against different target communities compared to the baselines. The authors measured GMB (i.e., the generalized mean of bias) which was introduced by the Google Conversation AI Team as part of their Kaggle competition[10]. The metric is expressed as $M_p(m_s) = \left(\frac{1}{N}\sum_s^N m_s^p\right)^{\frac{1}{p}}$ where m_s is the bias metric for the target group s, N is the total number of subgroups and M_p is the pth power mean function. The value of m_s is calculated in three different ways—(i) subgroup AUC (those toxic and normal posts are chosen from the test set that mentions the community under consideration and the ROC-AUC score is computed for this set), (ii) BPSN (background positive, subgroup negative) AUC (the normal posts that mention the community and the toxic posts that do not mention the community are chosen from the test set and the ROC-AUC score is computed for this set), and BNSP (background negative, subgroup positive) AUC (the toxic posts that mention the community and the normal posts that do not mention the community are chosen from the test set and the ROC-AUC score is computed for this set). As is apparent from the values of these different bias metrics in table 4.7, the BERT-HateXplain model is less biased than the standard baselines.

4.3.2.4 Mitigation of hate speech
Once hate speech or similar harmful content is detected on a social media platform, the most common actions that are taken for moderation are (i) deletion of posts, (ii) suspension of user accounts, or (iii) shadow banning whereby the posts of a user or the user themself is banned from different parts of an online community without the user realizing that they are banned.

---

[10] https://www.kaggle.com/c/jigsaw-unintended-bias-intoxicity-classification/overview/evaluation

Effectiveness of banning: one of the best case studies on the effectiveness of banning was done by [126] on the dataset pertaining to Reddit ban that took place in 2015. In 2015, Reddit closed several subreddits due to violations of Reddit's anti-harassment policy. Foremost among them were r/fatpeoplehate and r/CoonTown. At the user level, following Reddit's 2015 ban, a significant fraction of the users from the banned communities left Reddit. Others migrated to other subreddits where hate was a prominent topic. At the community level, an interesting observation is that the migrant users did not bring hate speech with them to their new communities, nor did the longtime residents pick it up from them. In other words, Reddit did not spread the 'hate infection'. Another observation is that users who got banned on Twitter/Reddit exhibited an increased level of activity and toxicity on Gab, although the audience they could potentially reach decreased [127].

Counterspeech: the *counterspeech* doctrine posits that the proper response to negative speech is to counter it back with *positive expression*. Combating hate speech in this way has some advantages: (i) it is faster, (ii) more flexible and responsive, capable of dealing with extremism from anywhere and in any language, and (iii) it does not form a barrier against the principle of free and open public space for debate. An example of a hate speech—counterspeech pair is presented in figure 4.5.

Types of counterspeech: the authors in [128] demonstrated that a monolithic form of counterspeech is usually not effective for different forms of hate speech. The authors further identified that typically a counterspeech can be of one of the following types—(a) presenting facts to correct misstatements or mis-perceptions—the counter speakers might use fact checkers and other pieces of evidence to prove that the hate content aimed against a target community is baseless and untrue, (b) pointing out hypocrisy/contradictions—the counter speakers can use methods to point out the hypocrisy in the hateful post (c) warning of online/offline consequences—the counter speakers can warn the hate speaker of consequences like banning/suspension of their account or legal actions against them, (d) affiliation—the counter speaker affiliates with the original speaker to establish a connection and, thereby, supports the victim, (e) visual communication—since images/videos are usually more persuasive than text, counter speakers often use them in the form of memes and video clips to combat hateful content, (f) humor and sarcasm—the counter speaker uses humorous counterspeech to de-escalate the conflict, reduce the harshness and aggressiveness of the hateful post and sometimes also drift the discussion, (g) denouncing hateful or dangerous speech—the counter speakers denounce the original hateful post and also at times explain why the post is dangerous, and (h)

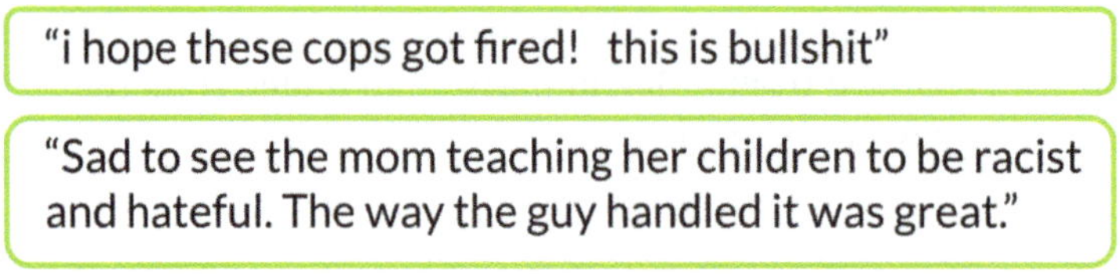

"i hope these cops got fired! this is bullshit"

"Sad to see the mom teaching her children to be racist and hateful. The way the guy handled it was great."

Figure 4.5. An example of hate speech (top) counterspeech (bottom) pair.

tone—the counter speakers can use a hostile tone to rebut the hate content (usually discouraged).

Counterspeech analysis: one of the early works was reported in [129]. In this work, the authors considered a set of hateful YouTube videos and scraped the comments. The target communities considered were Jews, African-Americans (black), and LGBT. Next, they annotated these comments in two stages. In the first stage, they annotated if a comment was a valid counterspeech or not. In the next stage, they annotated the comments marked as counterspeech in the first stage with one or more of the types. At the end of the first step, the authors obtained 6898 counterspeech comments and 7026 non-counterspeech comments. The annotations of the types of the 6898 counterspeech comments resulted in the distribution shown in table 4.8. The table reveals that for all three targets, users often post hostile comments in order to counter the hateful content of the video. While this is a usually discouraged strategy, it seems to be quite abundantly used. Apart from that, one observes that for different target communities, the counterspeech types that are used more are different. For instance, for the hateful videos against Jews, denouncing, positive tone, and presentation of facts seem to be used more. In the case of black people, denouncing and warning of online/offline consequences seem to be used more. Finally, for the LGBT community humor and pointing out hypocrisy are most predominantly used. The authors further estimated the engagement capability of the different types based on the number of likes and replies received by the counterspeech comments falling in those types. They found affiliation, denouncing of hateful comments, and warning of online/offline consequences to be the most effective ones for the black target videos. For the videos targeting Jews, the most effective types are affiliation and pointing out hypocrisy/contradiction. For the videos targeting the LGBT community, the effective types are humor, presenting of facts, and pointing out hypocrisy/contradiction.

Counterspeech generation: the traditional approach for obtaining counterspeech has been to recruit experts from NGOs with wide experience in framing such

Table 4.8. Distribution of different types of counterspeech in the YouTube comments dataset.

	Target community			
Type of counterspeech	Jews	Black people	LGBT	Total
Presenting facts	308	85	359	752
Pointing hypocrisy/contradictions	282	230	526	1038
Warning of consequences	112	417	199	728
Affiliation	206	159	200	565
Denouncing	376	482	473	1331
Humor/sarcasm	227	255	618	1100
Positive tone	359	237	268	864
Hostile language	712	946	1083	2741
Total	2582	2811	3726	9119

counterspeech posts. As the counterspeech technique became popular in the online world, crowd workers also became suitable for framing such posts. Nevertheless, the efforts mostly remained manual posing several challenges like the difficulty of interpretation, difficulty in framing, psychological toll, etc. Of late, large natural language generation models have been shown to be very useful. A welcome step to reduce the burden of manual labor would be if these models could be fine-tuned to generate counterspeech automatically. However, such fine-tuning would require training data and some efforts have been made to develop such data as shown in table 4.9. All the dataset comprises hate-counterspeech pairs built in different ways. For [128], both the hate videos and the counterspeech comments were curated from YouTube and were annotated by a set of student volunteers. For the datasets released by [69] and [130] the hate speech data was curated online while the counterspeech data was crafted by the crowd and the NGO experts respectively. The paper in [131] presented an author-reviewer framework to obtain scalable counterspeech data. An author (typically a crowd worker) is tasked with counterspeech generation in response to a hate speech comment and a reviewer can be a human or a classifier model that filters the produced output. A validation/post-editing phase is conducted with NGO operators over the filtered data. This framework is scalable allowing us to obtain datasets that are suitable in terms of diversity, novelty, and quantity. Based on this framework the authors in [89] developed a sizeable multi-target counterspeech dataset. One of the notable works in this space is the generate-prune-select scheme proposed by the authors in [132]. In the generate step, the primary objective is to produce a diverse set of candidates for counterspeech selection. The authors extracted all available counterspeech data points present in the training data and used a generative model to expand this dataset. In particular, the authors used an RNN-based variational autoencoder [133], that incorporates the globally distributed latent representations of all sentences to generate candidates. The next step consists of pruning this expanded set to remove all grammatically incorrect sentences. To do this, the authors built a grammaticality classifier trained on the linguistic acceptability (CoLA) [134] corpus. In the final step, the authors attempted to select the most relevant responses. To this purpose, they modified a pretrained response selection model for task-oriented dialog systems [135] and fine-tuned it on their dataset. The authors showed that they outperformed the state-of-the-art generation methods across multiple datasets in terms of diversity, relevance, and language quality. The authors also evaluated the outputs through human judges and obtained the same results, i.e., their model outperformed the baselines in terms of all three metrics—diversity, relevance, and

Table 4.9. Different types of counterspeech dataset. HS: hate speech, CS: counterspeech.

Type	HS source	CS source	Annotation	Annotators
Crawling [58]	Online	Online	Labeling	Crowd
Crowdsourcing [69]	Online	Synthetic	Response generation	Crowd
Niche sourcing [130]	Online/ Synthetic	Synthetic	Response generation	Experts

language quality. With such data in place, recently large transformer-based language generation models, e.g., GPT, T5 etc, could be used to automatically produce counterspeech. The authors in [136] used a DialoGPT model to generate counterspeech corresponding to a particular hate speech and subsequently used an ensemble of generative discriminators (GEDI) to make the counterspeech more polite, detoxified, and emotionally laden. In other words, the authors used a decoding time algorithm to control the output of the counterspeech generation model. A class-conditioned language model is constrained to have a desired control code (e.g., polite, non-toxic, etc) and an undesired control code (e.g., non-polite, toxic, etc). Using this model the authors showed that the politeness and the detoxification scores increased by around 15% and 6% respectively, while the overall emotion in the counterspeech increased by at least 10% across all the datasets.

4.3.3 Media bias

Media bias refers to the bias introduced by journalists and news producers in the media industry by a disproportionate selection of various events and stories that they report and the coverage of such events/stories thereof. Journalism is guided by the following *five* values.

- *Honesty*: journalists should report true news to the best of their knowledge. They should not twist/obfuscate/fake news content that mediates a wrong impression in the public.
- *Independence and objectivity*: journalists should not present opinions but facts (and only facts) in their reports. Any subject matter impacting the personal benefits of a journalist should not be covered as this could permeate bias in the coverage. In cases where a journalist has financial/personal interests, this needs to be disclosed.
- *Fairness*: the facts presented by journalists should be neutral and impartial and cover all sides of the story in a fair and unbiased way.
- *Diligence*: a journalist should do the due diligence to gather the news in such a way that it gives an appropriate understanding of the subject matter.
- *Accountability*: a journalist must take responsibility for the news items they had covered and be open to criticism.

Any deviation from the above might be a potential source of media bias. The issue of fairness and balance in the press has been an active area of research for at least the last five decades. A survey conducted in 1977 among a group of 746 editors showed that accuracy and impartiality of reporting are the most important tenets of editorial quality. Another survey conducted among 172 editors, educators, and national leaders found that 97% of the respondents felt that newspapers should endeavor to be fair and balanced in their reporting. In the following, we will look at some of the key research works that have been undertaken in this domain from both a journalistic and, a computational point of view.

Fairness and balance in the prestige press: the authors in [137] compared nine prestige newspapers with 12 large circulation newspapers that are typically less prestigious in terms of their likelihood of covering both sides of a community controversy, thereby, presenting more balanced news. The authors identified the prestige newspapers based on trade journals and previous research. Examples of prestige newspapers included New York Times and Washington Post. The authors obtained subscriptions for these 21 newspapers for a period of one week—April 21–26, 1986. Staff stories narrating a controversy and relating to either the (i) local government, (ii) the local public school education, or (iii) the local business were identified by two annotators. A third annotator adjudicated those judgments to take the final decision as to whether to include or exclude the stories. This resulted in a total of 343 stories based on which the authors conducted their analysis. The author presented two different analyses—(a) fairness and (b) balance. Fairness was defined as whether the two parties involved in a controversy were either (a) successfully contacted or at least a contact attempt was mentioned, or (b) not contacted at all. Balance was determined by the number of words given to each side of the controversy; in order to accommodate variations in type height, type density, and column width, a words-per-inch formula was used. The authors counted the number of the most prominent entities on either side of the controversy as well as the supporting sources. Finally, balance was defined as the average difference in the percentage of the total story in terms of the words-per-inch given to each side. The key results from this study are reproduced in table 4.10. Prestige press newspapers are better in terms of both fairness and balance than large

Table 4.10. Fairness (F) and balance (B) based ranks of the prestige and large circulation newspapers. Cells marked in *green* represent the best newspaper in terms of a metric (F or B) and cells marked in *red* represent the worst.

Prestige newspapers	% stories w/o both sides contacted	F Rank	Avg. diff. in % of story	B Rank	Mean Rank	N
Los Angeles Times	0.0	1	22.6	4	2.5	13
The (Louisville) Courier-Journal	0.0	1	25.4	9	5.0	10
The New York Times	7.1	7	20.7	1	4.0	14
The Washington Post	5.6	5	20.8	2	3.5	18
High circulation newspapers						
New York Daily News	45.5	21	43.3	21	21.0	11
The Detroit News	25.0	16	37.3	20	18.0	12
Chicago Sun-Times	36.0	20	33.2	17	18.5	25
Newsday	5.6	5	27.9	12	8.5	18
San Francisco Chronicle	12.5	10	23.8	6	8.0	24
Minneapolis Star & Tribune	0.0	1	26.5	10	5.5	13

circulation newspapers. Among the prestige press newspapers, Los Angeles Times and The Courier-Journal are the best in terms of fairness and The New York Times and The Washington Post are best in terms of balance. Among the large circulation newspapers the Minneapolis Star and Tribune is best in terms of fairness and San Francisco Chronicle is best in terms of balance. The worst in both fairness and balance across the 21 newspapers is the New York Daily News. The Chicago Sun-Times and Detroit News are the second to worst in terms of fairness and balance respectively. Further, the authors noted that there is a larger variance in fairness values across the newspapers compared to balance. In addition, the difference between fairness of the prestige press newspapers and large circulation newspapers was found to be larger compared to balance. A series of follow up works in this direction have investigated this problem from various perspectives culminating in a comprehensive book [138]. Finally, in [139] the authors demonstrated how native advertising hampers the quality of prestige media outlets.

Computational approaches: One of the early computational approaches was presented in [140] where the authors proposed NewsCube which, at that time, was a novel Internet service to mitigate media bias. The authors noted the far-reaching ramifications of media bias such as increased political polarization and eventual social conflict which constituted their primary motivations to develop NewsCube. Further, they correctly pointed out that the biases are introduced in such a nuanced fashion that most often it is impossible for the readers to penetrate into their authenticity. To circumvent this problem they built an automatic pipeline that creates and provides classified viewpoints about a news event in real time. Such plurality in viewpoints allows readers to see the event from different angles, thereby, helping them to construe a more balanced view of their own. The NewsCube system has three components as shown in figure 4.6. We now discuss the steps involved in its functioning.

- *Article collection*: This part is responsible for crawling news pages from news websites. The irrelevant content like ads, reader's comments, and metadata are removed retaining pure text. Next, parts of the news text are tagged by special structural information like ⟨ Title ⟩, ⟨ Subtitle ⟩, and keywords.

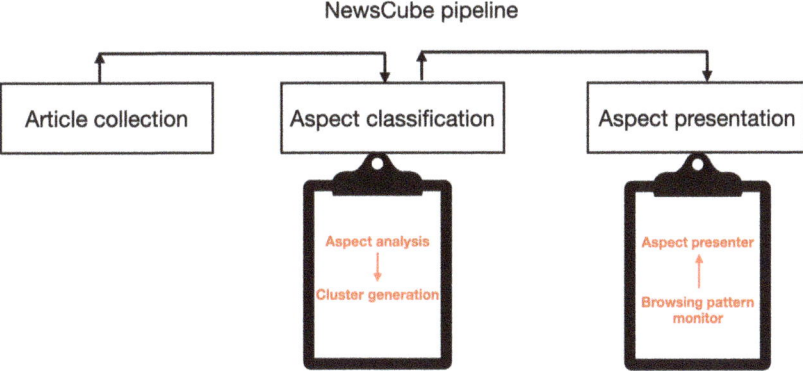

Figure 4.6. The overall architecture of the NewsCube system.

- *Aspect classification*: The authors used the structure-based method to extract the aspects. This method leverages on one of the famous news writing rules, called the 'inverted pyramid style of news writing'. According to this rule, journalists should arrange the facts in descending order of their importance while drafting an article. A representation of this structure is shown in figure 4.7. The article talks about mushrooms as a natural source of Vitamin D. The two example keywords—'mushroom' and 'vitamin D'—appear in different parts of the articles corresponding to the different segments of the inverted pyramid—head, sub-head, lead, and main text. The pyramidal structure has been used by the authors to select the aspect keywords as well as for the calculation of the weight of these keywords. The authors extracted keywords primarily from the core parts of the article, i.e., head, sub-head, and lead. The weight of the keywords is calculated based on the frequency as well as their location of occurrence in the main text. Further, the amount of text used to describe a keyword is also factored into the weight metric. Each article is finally represented as a vector of keywords with entries corresponding to the weights of these keywords. In order to compute the aspect groups the authors performed hierarchical agglomerative clustering. Starting with each individual article in a different cluster, the method merges the most similar pair of clusters in an iterative fashion. The similarity between a pair of articles is computed in terms of the cosine angle between their corresponding weight vectors.
- *Aspect presentation*: The NewsCube interface that the authors developed presents the news items in the different aspect clusters in which the entire content is grouped. Each aspect cluster is given equal space on the interface so as to increase the diversity of the user attention thus mitigating possible biases.

The authors conducted a survey among 33 participants for thenews event—'a protest against the US military base expansion'. The authors formed groups of two participants—two participants who were against the protest (negative group), two who supported the protesters (positive group), and two who did not have a preformed opinion (neutral group). In each group, one participant read the news through the NewsCube service and the other through the GoogleNews service. They

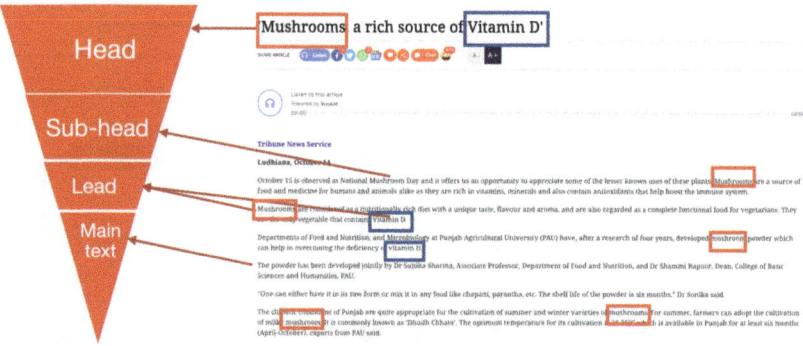

Figure 4.7. The inverted pyramid style of news composition.

were asked to use the service for 15 min. None of the participants were aware of the end goals of the NewsCube service. Subsequently, they were interviewed based on the following questions—(a) the demand felt by them to compare and contrast the multiple aspects present in a news event, (b) the diversity of the articles they read, and (c) the influence of the articles on their views. In response to the first question, 79% of the participants said that 'they (strongly) felt that it is necessary to compare multiple articles' in the past, (i.e., before knowing NewsCube). Further, 70% said that they 'actually compared articles once (or more) before'. In answer to the second question, 72% of the participants said that they '(strongly) felt like to read various articles with different aspects' while using the NewsCube service. 70% participants said that 'they did actually read various articles'. For the third question, 48% participants said that going through the different aspects by using the NewsCube service helped them to form a more balanced view of the news events.

Media bias monitor: In a recent study [141], the authors presented a novel service —the Media Bias Monitor—that is capable of quantifying the ideological biases of 20 448 news outlets on Facebook. To this purpose, the authors designed a crawler that makes use of the Facebook marketing API to gather information about a large number of media outlets present on Facebook. Next, they employed the Facebook audience API and collected different demographic information about the audience reached by these news outlets on Facebook. Along with information on the ideological leanings of the audience, the demographic information also consisted of other dimensions such as gender, age, national identity, racial affinity, and income level. They empirically demonstrated that the ideological leaning of a news source can be quite accurately captured by estimating the extent of over- or under-representation of the liberals/conservatives in the audience of that source. Some of the interesting observations noted by the authors were (a) for Fox News, 50% of the Facebook audience classified themselves as conservative out of which 28% called themselves very conservative, (b) for Breitbart, as high as 89% audience classified themselves as conservative (23% as very conservative), (c) for The New York Times, 57% audience classified themselves as liberal out of which around 30% called themselves very liberal, (d) some of the outlets dominated by the male audience were GuysGuns (94%), and SOFREP.com (92%), (e) outlets that were dominated by an audience who were both conservative and male were Drudge Report (63%) and Rush Limbaugh Show (60%), (f) outlets that were dominated by a group of audience who were both conservative and female were PrayAmerica (82%), Breaking Christian News (71%), and Lifenews.com (76%), and (g) The Man Repeller (96%), Feministing.com (91%), and Everyday Feminism (90%) were some of the outlets dominated by audiences who were both liberal and female.

Practice problems

Q1. You are appointed as a content governance expert at 'Convo' a social media platform. You have already implemented a **FakeDetect** model which detects fake news. The way the system works is all the posts are passed through the **FakeDetect** model which detects the post as fake news or not based on a threshold. The posts filtered by the model are further evaluated by the moderators who decide on the mitigation strategy for a post.

Note: For all the questions, you are not allowed to change the model **FakeDetect** but can use it as a black box and train or predict some samples.

  (a) Write the definition of fake news and describe how it is different from other related concepts like misinformation, disinformation, etc.
  (b) Whilst reviewing posts about the 2020 pandemic you found several reports of posts that were fake news but were not captured by **FakeDetect**. On carefully looking at them you found that these were mostly fake news about Chinese people. Your team decides to make your model better at detecting such posts and asks for your guidance. What changes will you make in the pipeline to achieve the same?
  - HINT: Try to consider the changes required in (i) data sampling, (ii) data annotation, and (iii) model evaluation.

Q2(a). What is clickbait and how are clickbait posts different from non-clickbait posts?

Q2(b). Suppose you want to create a dataset for fake news detection. When would you ideally choose expert-oriented fact checking and when would you choose crowdsourced fact checking?

You are scraping data from the Twitter profile containing the demographic information (i.e., age, race, gender) of the users. However, you are finding it difficult to collect this information directly from the user profiles of most of the individuals as they have not made this information public. Can you think of any alternative approach for gathering their demographic information from their social media profile? (Mention any assumption you are going to make; hint: think from the ML perspective).

Q2(c). Assume that you are analyzing user behavior on clickbait posts versus non-clickbait posts on Twitter. You attempt to measure how long the users remain engaged (retweet, comment, etc) with the clickbait and non-clickbait tweets from their time of production. Figure 4.8 shows the longevity of different tweets in the clickbait and non-clickbait categories. Discuss the differences in user engagement for

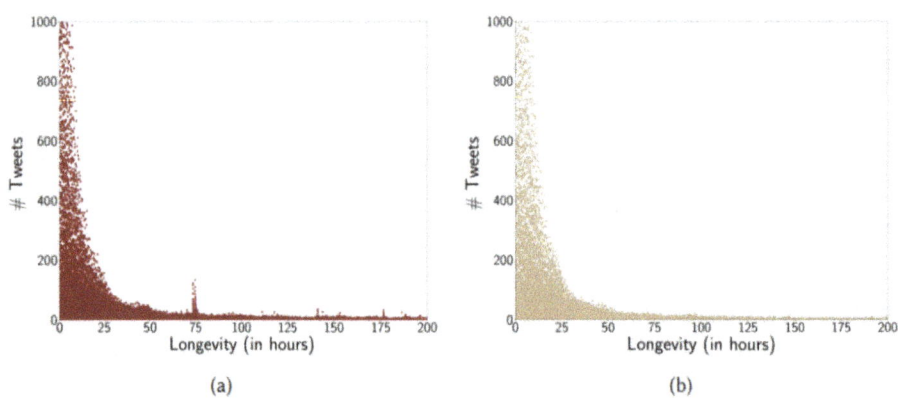

Figure 4.8. Scatter plots of longevity values (in hours) for various tweets in the (a) clickbait and (b) non-clickbait categories. Reproduced from [142] with permission from the Association for Computing Machinery.

the clickbait versus non-clickbait tweets. What do the sudden spikes in the longevity values of clickbait engagement indicate?

Q3. In the paper 'Spread of hate speech in online social media' [58] you noticed several interesting trends while comparing the cascade properties of the posts of hateful and non-hateful users. A similar study was performed on the Twitter platform, the plot in figure 4.9 shows kernel density estimations (see **Note** below) of joining times of different categories of users. **Hateful users** refer to users posting hate speech and **normal users** are users who never post hate speech. **Hateful/normal neighbors** refer to the direct followers of a hateful/normal user. **Suspended users** are users who are banned while **active users** refer to all the users who are not suspended and have posted at least once. Answer the following questions:

(a) What is the main observation from this plot?
(b) What does this observation signify?
(c) Do you expect the same behavior in the Gab social network? Give justification.

Note: A kernel density estimate (KDE) plot is a method for visualizing the distribution of observations in a dataset, analogous to a histogram. KDE represents the data using a continuous probability density curve in one or more dimensions.

Q4. For the network shown in figure 4.10, deduce the transition matrix P from its adjacency matrix using $\alpha = 0.8$. Assuming the exposure of a node is the probability that the same can be reached from any given node on the graph, find out the exposure of all the nodes based on your transition probability matrix. (Use the instructions given on Page 466 of [143] for deducing the transition probability matrix.)

Q5. You are working on building a knowledge graph and related queries. You have been tasked with evaluating the knowledge graph creation pipeline and evaluating

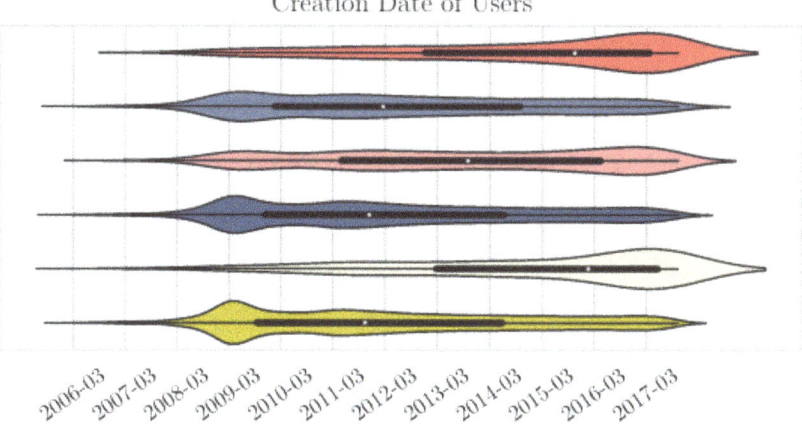

Figure 4.9. KDEs of the creation dates of user accounts. The white dot indicates the median and the thicker bar indicates the first and third quartiles. Reproduced with permission from [64]. Copyright 2018, Association for the Advancement of Artificial Intelligence, all rights reserved.

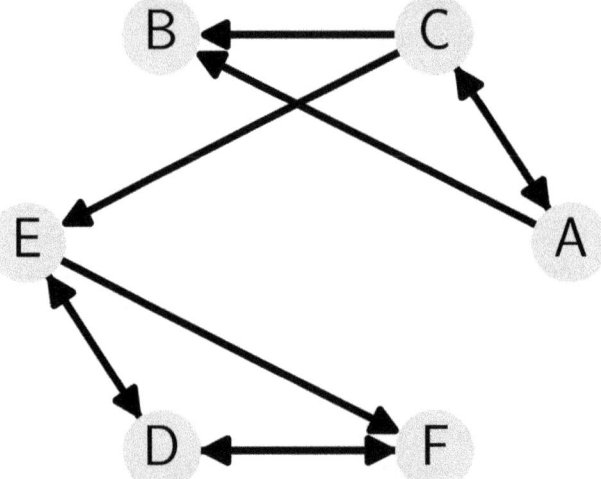

Figure 4.10. Directed graph for Q4.

different pre-decided queries. In order to do this you wrote a dummy text as noted below:

> Lily is interested in the Da Vinci who painted the painting of Mona Lisa. The painting is located in Louvre Museum, Paris. Lily is a friend of James who also likes the painting of Mona Lisa and has visited the Louvre Museum, Paris. James is staying at Tour Eiffel, Paris. Paris is a place. Da Vinci, Lily and James are human beings.

(a) Draw the knowledge graph using the above unstructured text by first creating $\langle S, P, O \rangle$ triplets and then converting them into a graph.

(b) Using the knowledge graph generated in the former question and the algorithm given in algorithm 1 find out whether the facts given below are true or not—
 (i) $\langle$Mona Lisa, is in, Tour Eiffel$\rangle$; follow open-world assumption (OWA)
 (ii) $\langle$Lily, painted, Mona Lisa$\rangle$; follow closed-world assumption (CWA)
 (iii) $\langle$Louvre, is located in, France$\rangle$; follow local closed-world assumption (LCWA)
 (iv) $\langle$James, is located in, Australia$\rangle$; follow local closed-world assumption (LCWA)

(c) Apart from these direct queries, you also want to evaluate semantic proximity function in their implementation. Semantic proximity between two entities s and o is defined in figure 4.11. Here, $k(v_i)$ denotes the degree of the particular

$$\mathcal{W}(P_{s,o}) = \mathcal{W}(v_1 \ldots v_n) = \left[1 + \sum_{i=2}^{n-1} \log k(v_i)\right]^{-1}$$

Figure 4.11. Equation for semantic proximity.

node v_i, i.e., the number of WKG statements it participates in. Calculate the semantic proximity for the following entity pairs—
(i) s = Lily and o = Paris
(ii) s = Da Vinci and o = Paris
- NOTE: If there are multiple paths between both entities, calculate the semantic proximity for all the paths.
- NOTE: The knowledge graph should be created from the available sentences. The algorithm does not have access to any other knowledge apart from this knowledge graph.

References

[1] Ali M, Sapiezynski P, Bogen M, Korolova A, Mislove A and Rieke A 2019 Discrimination through optimization: how Facebook's ad delivery can lead to biased outcomes *Proc. ACM on Human–Computer Interaction* vol 3 pp 1–30
[2] Mehrotra R, Anderson A, Diaz F, Sharma A, Wallach H and Yilmaz E 2017 Auditing search engines for differential satisfaction across demographics *Proc. 26th Int. Conf. on World Wide Web Companion* pp 626–33
[3] Kulshrestha J, Eslami M, Messias J, Zafar M B, Ghosh S, Gummadi K P and Karahalios K 2017 Quantifying search bias: investigating sources of bias for political searches in social media *Proc. 2017 ACM Conf. on Computer Supported Cooperative Work and Social Computing* pp 417–32
[4] Robertson R E, Jiang S, Joseph K, Friedland L, Lazer D and Wilson C 2018 Auditing partisan audience bias within google search *Proc. ACM on Human–Computer Interaction* vol 2 CSCW, pp 1–22
[5] Pariser E 2011 *The Filter Bubble: What the Internet Is Hiding from You* (London: Penguin)
[6] Ribeiro M H, Ottoni R, West R, Almeida V A F and Meira W Jr 2020 Auditing radicalization pathways on YouTube *Proc. 2020 Conf. on Fairness, Accountability, and Transparency* pp 131–41
[7] Dash A, Chakraborty A, Ghosh S, Mukherjee A and Gummadi K P 2021 When the umpire is also a player: bias in private label product recommendations on e-commerce marketplaces *Proc. 2021 ACM Conf. on Fairness, Accountability, and Transparency* pp 873–84
[8] Press Information Bureau 2018 Review of Policy on Foreign Direct Investment (FDI) in E-commerce http://www.pib.nic.in/PressReleseDetail.aspx?PRID=1557380
[9] Kalra A 2021 *Amazon Documents Reveal Company's Secret Strategy to Dodge India's Regulators* https://reut.rs/2Yt3GlL
[10] USA House Committee on the Judiciary 2020 *Judiciary Antitrust Subcommittee Investigation Reveals Digital Economy Highly Concentrated, Impacted by Monopoly Power* https://judiciary.house.gov/news/documentsingle.aspx?DocumentID=3429

[11] Antitrust Subcommittee 2020 *Investigation of Competition in Digital Markets* https://judiciary.house.gov/uploadedfiles/competition_in_digital_markets.pdf
[12] Sarkar S, Bhowmick S and Mukherjee A 2018 On rich clubs of path-based centralities in networks *Proc. 27th ACM Int. Conf. on Information and Knowledge Management*
[13] Sarkar S, Sikdar S, Bhowmick S and Mukherjee A 2018 Using core-periphery structure to predict high centrality nodes in time-varying networks *Data Min. Knowl. Discov.* **32** 1369–96
[14] EU Press Release 2020 *Antitrust: Commission Sends Statement of Objections to Amazon for the Use of Non-Public Independent Seller Data and Opens Second Investigation into Its E-commerce Business Practices* https://judiciary.house.gov/news/documentsingle.aspx?DocumentID=3429
[15] Domm P 2013 False Rumor of Explosion at White House Causes Stocks to Briefly Plunge; AP Confirms Its Twitter Feed Was Hacked https://www.cnbc.com/id/100646197
[16] Guterres A 2019 *United Nations Strategy and Plan of Action on Hate Speech* https://www.un.org/en/genocideprevention/hate-speech-strategy.shtml
[17] Ryan D et al 2020 *AI Advances to Better Detect Hate Speech* https://ai.facebook.com/blog/ai-advances-to-better-detect-hate-speech/
[18] Böhm C, De Melo G, Naumann F and Weikum G 2012 LINDA: distributed web-of-data-scale entity matching *Proc. 21st ACM Int. Conf. on Information and Knowledge Management* pp 2014–8
[19] Whang S E, Marmaros D and Garcia-Molina H 2012 Pay-as-you-go entity resolution *IEEE Trans. Knowl. Data Eng.* **25** 1111–24
[20] Karapiperis D, Gkoulalas-Divanis A and Verykios V S 2018 Summarization algorithms for record linkage *Proc. 21st Int. Conf. on Extending Database Technology, EDBT 2018* pp 73–84
[21] Dong X, Halevy A and Madhavan J 2005 Reference reconciliation in complex information spaces *Proc. 2005 ACM SIGMOD Int. Conf. on Management of Data* pp 85–96
[22] Weis M and Naumann F 2006 Detecting duplicates in complex xml data *22nd Int. Conf. on Data Engineering (ICDE'06)* (Piscataway, NJ: IEEE) p 109
[23] Ioannou E, Niederée C and Nejdl W 2008 Probabilistic entity linkage for heterogeneous information spaces *Int. Conf. on Advanced Information Systems Engineering* (Berlin: Springer) pp 556–70
[24] Miller K, Jordan M and Griffiths T 2009 Nonparametric latent feature models for link prediction *Adv. Neural. Inf. Process. Syst.* **22** https://proceedings.neurips.cc/paper/2009/file/437d7d1d97917cd627a34a6a0fb41136-Paper.pdf
[25] Toutanova K, Chen D, Pantel P, Poon H, Choudhury P and Gamon M 2015 Representing text for joint embedding of text and knowledge bases *Proc. 2015 Conf. on Empirical Methods in Natural Language Processing* pp 1499–509
[26] Toutanova K and Chen D 2015 Observed versus latent features for knowledge base and text inference *Proc. 3rd Workshop on Continuous Vector Space Models and Their Compositionality* pp 57–66
[27] Quattoni A, Collins M and Darrell T 2004 Conditional random fields for object recognition *Advances in Neural Information Processing Systems 17 (NIPS 2004)* https://proceedings.neurips.cc/paper/2004/file/0c215f194276000be6a6df6528067151-Paper.pdf
[28] Lafferty J D, McCallum A and Pereira F C N 2001 Conditional random fields: probabilistic models for segmenting and labeling sequence data *Proc. 18th Int. Conf. on Machine Learning* (San Francisco, CA: Morgan Kaufmann Publishers) pp 282–9

[29] Undeutsch U 1967 Beurteilung der glaubhaftigkeit von aussagen *Handbuch der Psychologie* vol 11 pp 26–181
[30] Zhou X, Jain A, Phoha V V and Zafarani R 2019 Fake news early detection: a theory driven model *Dig. Threats: Res. Prac.* **1** 1–25
[31] Recasens M, Danescu-Niculescu-Mizil C and Jurafsky D 2013 Linguistic models for analyzing and detecting biased language *Proc. 51st Annual Meeting of the Association for Computational Linguistics (vol 1: Long Papers)* pp 1650–9 https://aclanthology.org/P13-1162.pdf
[32] Ji Y and Eisenstein J 2014 Representation learning for text-level discourse parsing *Proc. 52nd Annual Meeting of the Association for Computational Linguistics (vol 1: Long Papers)* pp 13–24 (Piscataway, NJ: IEEE) https://aclanthology.org/P14-1002.pdf
[33] Wu K, Yang S and Zhu K Q 2015 False rumors detection on sina weibo by propagation structures *2015 IEEE 31st Int. Conf. on Data Engineering* (Piscataway, NJ: IEEE) pp 651–62
[34] Liu Y and Wu Y-F 2018 Early detection of fake news on social media through propagation path classification with recurrent and convolutional networks *Proc. AAAI Conf. on Artificial Intelligence* vol 32
[35] Gupta M, Zhao P and Han J 2012 Evaluating event credibility on Twitter *Proc. 2012 SIAM Int. Conf. on Data Mining* (Philadelphia, PA: SIAM) pp 153–64
[36] Gupta S, Thirukovalluru R, Sinha M and Mannarswamy S 2018 CIMTDetect: a community infused matrix-tensor coupled factorization based method for fake news detection *IEEE/ACM Int. Conf. on Advances in Social Networks Analysis and Mining (ASONAM)* (Piscataway, NJ: IEEE) pp 278–81
[37] Shu K, Wang S and Liu H 2019 Beyond news contents: the role of social context for fake news detection *Proc. 12th ACM Int. Conf. on Web Search and Data Mining* pp 312–20
[38] Ruchansky N, Seo S and Liu Y 2017 CSI: a hybrid deep model for fake news detection *Proc. 2017 ACM on Conf. on Information and Knowledge Management* pp 797–806
[39] Zhang J, Cui L, Fu Y and Gouza F B 2018 *Fake News Detection with Deep Diffusive Network Model* (arXiv:1805.08751v2 [cs.SI])
[40] Jin Z, Cao J, Zhang Y and Luo J 2016 News verification by exploiting conflicting social viewpoints in microblogs *Proc. AAAI Conf. on Artificial Intelligence* vol 30
[41] Shu K, Sliva A, Wang S, Tang J and Liu H 2017 Fake news detection on social media: a data mining perspective *ACM SIGKDD Explor. Newsl.* **19** 22–36
[42] Shu K, Mahudeswaran D, Wang S, Lee D and Liu H 2018 *Fakenewsnet: A Data Repository with News Content, Social Context and Dynamic Unformation for Studying Fake News on Social Media* (arXiv:1809.01286v3 [cs.SI])
[43] Vo N and Lee K 2020 *Where are the Facts? Searching for Fact-Checked Information to Alleviate the Spread of Fake News* (arXiv:2010.03159v1 [cs.IR])
[44] FactCheck 2022 *Factcheck.org: A Project of the Annenberg Public Policy Center* https://www.factcheck.org/
[45] Wang W Y 2017 'Liar, liar pants on fire': a new benchmark dataset for fake news detection *Proc. 55th Annual Meeting of the Association for Computational Linguistics (vol 2: Short Papers)* (Piscataway, NJ: IEEE) pp 422–6
[46] Micallef N, He B, Kumar S, Ahamad M and Memon N 2020 The role of the crowd in countering misinformation: a case study of the COVID-19 infodemic *2020 IEEE Int. Conf. on Big Data (Big Data)* (Piscataway, NJ: IEEE) pp 748–57

[47] Nakamura K, Levy S and Wang W Y 2019 Fakeddit: a new multimodal benchmark dataset for fine-grained fake news detection *Proc. of the Twelfth Language Resources and Evaluation Conf. (Marseille)* pp 6149–57

[48] Nan Q, Cao J, Zhu Y, Wang Y and Li J 2021 MDFEND: multi-domain fake news detection *Proc. 30th ACM Int. Conf. on Information and Knowledge Management* pp 3343–7

[49] Nielsen D S and McConville R 2022 MuMiN: a large-scale multilingual multimodal fact-checked misinformation social network dataset *Proc. 45th Int. ACM SIGIR Conf. on Research and Development in Information Retrieval (SIGIR)* (New York: ACM)

[50] TruthOrFiction 2022 Truthorfiction: Seeking Truth and Exposing Fiction Since 1999 https://www.truthorfiction.com/

[51] ElSherief M, Nilizadeh S, Nguyen D, Vigna G and Belding E 2018 Peer to peer hate: hate speech instigators and their targets *Proc. Int. AAAI Conf. on Web and Social Media* vol 12

[52] Kwok I and Wang Y 2013 Locate the hate: detecting tweets against blacks *27th AAAI Conf. on Artificial Intelligence*

[53] Farrell T, Fernandez M, Novotny J and Alani H 2019 Exploring misogyny across the manosphere in reddit *Proc. 10th ACM Conf. on Web Science* 87–96

[54] Tahmasbi F, Schild L, Ling C, Blackburn J, Stringhini G, Zhang Y and Zannettou S 2021 'Go eat a bat, chang!': on the emergence of sinophobic behavior on web communities in the face of COVID-19 *Proc. Web Conf. 2021* pp 1122–33

[55] Mittos A, Zannettou S, Blackburn J and De Cristofaro E 2020 'And we will fight for our race!' A measurement study of genetic testing conversations on reddit and 4chan *Proc. Int. AAAI Conf. on Web and Social Media* vol 14 pp 452–63

[56] Sipka A, Hannak A and Urman A 2022 Comparing the language of QAnon-related content on Parler, Gab, and Twitter *14th ACM Web Science Conf. 2022* pp 411–21

[57] Zannettou S, Bradlyn B, De Cristofaro E, Kwak H, Sirivianos M, Stringini G and Blackburn J 2018 What is gab: a bastion of free speech or an alt-right echo chamber *Companion Proc. Web Conf. 2018* pp 1007–14

[58] Mathew B, Dutt R, Goyal P and Mukherjee A 2019 Spread of hate speech in online social media *Proc. 10th ACM Conf. on Web Science* pp 173–82

[59] Golub B and Jackson M O 2010 Naive learning in social networks and the wisdom of crowds *Am. Econ. J. Microecon.* **2** 112–49

[60] Bakshy E, Hofman J M, Mason W A and Watts D J 2011 Everyone's an influencer: quantifying influence on Twitter *Proc. 4th ACM Int. Conf. on Web Search and Data Mining* pp 65–74

[61] Vosoughi S, Roy D and Aral S 2018 The spread of true and false news online *Science* **359** 1146–51

[62] Mathew B, Illendula A, Saha P, Sarkar S, Goyal P and Mukherjee A 2020 Hate begets hate: a temporal study of hate speech *Proc. ACM on Human–Computer Interaction* vol 4 CSCW2, pp 1–24

[63] Papasavva A, Zannettou S, De Cristofaro E, Stringhini G and Blackburn J 2020 Raiders of the lost Kek: 3.5 years of augmented 4chan posts from the politically incorrect board *Proc. Int. AAAI Conf. on Web and Social Media* 14 pp 885–94

[64] Ribeiro M H, Calais P H, Santos Y A, Almeida V A F and Meira W Jr 2018 Characterizing and detecting hateful users on Twitter *12th Int. AAAI Conf. on Web and Social Media*

[65] Saha P, Mathew B, Garimella K and Mukherjee A 2021 'Short is the road that leads from fear to hate': fear speech in Indian WhatsApp groups *Proc. Web Conf.* **2021** 1110–21

[66] Albadi N, Kurdi M and Mishra S 2018 Are they our brothers? Analysis and detection of religious hate speech in the Arabic Twittersphere *2018 IEEE/ACM Int. Conf. on Advances in Social Networks Analysis and Mining (ASONAM)* (Piscataway, NJ: IEEE) pp 69–76

[67] Mollas I, Chrysopoulou Z, Karlos S and Tsoumakas G 2020 *ETHOS: An Online Hate Speech Detection Dataset* (arXiv:2006.08328 [cs.CL])

[68] Kirk H R, Vidgen B, Röttger P, Thrush T and Hale S A 2021 *Hatemoji: A Test Suite and Adversarially-Generated Dataset for Benchmarking and Detecting Emoji-Based Hate* (arXiv:2108.05921 [cs.CL])

[69] Qian J, Bethke A, Liu Y, Belding E and Wang W Y 2019 *A benchmark Dataset for Learning to Intervene in Online Hate Speech* (arXiv:1909.04251 [cs.CL])

[70] Kennedy B *et al* 2018 The Gab Hate Corpus: A Collection of 27k Posts Annotated for Hate Speech *PsyArXiv* https://psyarxiv.com/hqjxn/

[71] Fersini E, Rosso P and Anzovino M 2018 Overview of the task on automatic misogyny identification at IberEval *Proc. 3rd Workshop on Evaluation of Human Language Technologies for Iberian Languages (IberEval 2018)* vol 2150 pp 214–28 https://ceur-ws.org/Vol-2150/overview-AMI.pdf

[72] Pamungkas E W, Basile V and Patti V 2020 Do you really want to hurt me? Predicting abusive swearing in social media *The 12th Language Resources and Evaluation Conf.* (Paris: European Language Resources Association) pp 6237–46 https://aclanthology.org/2020.lrec-1.765

[73] Suryawanshi S, Chakravarthi B R, Arcan M and Buitelaar P 2020 Multimodal meme dataset (multioff) for identifying offensive content in image and text *Proc. 2nd Workshop on Trolling, Aggression and Cyberbullying* pp 32–41 https://aclanthology.org/2020.trac-1.6

[74] Jiang A, Yang X, Liu Y and Zubiaga A 2022 SWSR: a chinese dataset and lexicon for online sexism detection *Online Soc. Netw. Media* **27** 100182

[75] Salminen J, Almerekhi H, Milenković M, Jung S-G, An J, Kwak H and Jansen B J 2018 Anatomy of online hate: developing a taxonomy and machine learning models for identifying and classifying hate in online news media *12th Int. AAAI Conf. on Web and Social Media*

[76] Mathew B, Saha P, Yimam S M, Biemann C, Goyal P and Mukherjee A 2021 Hatexplain: a benchmark dataset for explainable hate speech detection *Proc. AAAI Conf. on Artificial Intelligence* vol 35 pp 14867–75

[77] Curry A C, Abercrombie G and Rieser V 2021 ConvAbuse: data, analysis, and benchmarks for nuanced detection in conversational AI *Proc. 2021 Conf. on Empirical Methods in Natural Language Processing* pp 7388–403

[78] Waseem Z 2016 Are you a racist or am I seeing things? Annotator influence on hate speech detection on Twitter *Proc. 1st Workshop on NLP and Computational Social Science* pp 138–42

[79] Davidson T, Warmsley D, Macy M and Weber I 2017 Automated hate speech detection and the problem of offensive language *Proc. Int. AAAI Conf. on Web and Social Media* pp 512–5

[80] Fortuna P and Nunes S 2018 A survey on automatic detection of hate speech in text *ACM Comput. Surv. (CSUR)* **51** 1–30

[81] Vasantharajan C and Thayasivam U 2021 Hypers@DravidianLangTech-EACL2021: offensive language identification in Dravidian code-mixed YouTube comments and posts

Proc. 1st Workshop on Speech and Language Technologies for Dravidian Languages pp 195–202 https://aclanthology.org/2021.dravidianlangtech-1.26

[82] Chakravarthi B R, Priyadharshini R, Ponnusamy R, Kumaresan P K, Sampath K, Thenmozhi D, Thangasamy S, Nallathambi R and McCrae J P 2021 Dataset for Identification of Homophobia and Transophobia in Multilingual YouTube Comments (arXiv:2109.00227 [cs.CL])

[83] Del Vigna F, Cimino A, Dell'Orletta F, Petrocchi M and Tesconi M 2017 Hate me, hate me not: hate speech detection on Facebook *Proc. 1st Italian Conf. on Cybersecurity (ITASEC17)* pp 86–95 https://ceur-ws.org/Vol-1816/paper-09.pdf

[84] Kiela D, Firooz H, Mohan A, Goswami V, Singh A, Ringshia P and Testuggine D 2020 The hateful memes challenge: detecting hate speech in multimodal memes *Advances in Neural Information Processing Systems 33: Annual Conf. on Neural Information Processing Systems 2020, NeurIPS 2020, December 6–12, 2020, Virtual*; H Larochelle, M Ranzato, R Hadsell, M-F Balcan and H-T Lin https://proceedings.neurips.cc/paper/2020/file/1b84c4cee2b8b3d823b30e2d604b1878-Paper.pdf

[85] Israeli A and Tsur O 2022 Free speech or free hate speech? Analyzing the proliferation of hate speech in parler *Proc. 6th Workshop on Online Abuse and Harms (WOAH)* pp 109–21

[86] Mekacher A and Papasavva A 2022 'I can't keep it up.' A dataset from the defunct Voat.co news aggregator *Proc. Int. AAAI Conf. on Web and Social Media* vol 16 1302–11 https://ojs.aaai.org/index.php/ICWSM/article/view/19382/19154

[87] Rieger D, Kümpel A S, Wich M, Kiening T and Groh G 2021 Assessing the extent and types of hate speech in fringe communities: a case study of alt-right communities on 8chan, 4chan, and reddit *Soc. Media Soc.* 7 20563051211052906

[88] Trujillo M Z, Gruppi M, Buntain C and Horne B D 2022 The MeLa BitChute dataset *Proc. Int. AAAI Conf. on Web and Social Media* **vol 16** pp 1342–51

[89] Fanton M, Bonaldi H, Tekiroğlu S S and Guerini M 2021 Human-in-the-loop for data collection: a multi-target counter narrative dataset to fight online hate speech *Proc. 59th Annual Meeting of the Association for Computational Linguistics and the 11th Int. Joint Conf. on Natural Language Processing (vol 1: Long Papers)* pp 3226–40

[90] Vidgen B, Thrush T, Waseem Z and Kiela D 2021 Learning from the worst: dynamically generated datasets to improve online hate detection *Proc. 59th Annual Meeting of the Association for Computational Linguistics and the 11th Int. Joint Conf. on Natural Language Processing (vol 1: Long Papers)* pp 1667–82

[91] Mulki H, Haddad H, Ali C B and Alshabani H 2019 L-HSAB: a Levantine Twitter dataset for hate speech and abusive language *Proc. 3rd Workshop on Abusive Language Online* pp 111–8

[92] Ousidhoum N, Lin Z, Zhang H, Song Y and Yeung D-Y 2019 *Multilingual and Multi-Aspect Hate Speech Analysis* (arXiv:1908.11049 [cs.CL])

[93] Caselli T, Novielli N, Patti V and Rosso P 2018 Sixth evaluation campaign of natural language processing and speech tools for Italian *EVALITA 2018. CEUR Workshop Proc. (CEUR-WS. org)* https://ceur-ws.org/Vol-2263/

[94] Sanguinetti M, Poletto F, Bosco C, Patti V and Stranisci M 2018 An Italian Twitter corpus of hate speech against immigrants *Proc. 11th Int. Conf. on Language Resources and Evaluation (LREC 2018)* https://aclanthology.org/L18-1443

[95] Garibo i Orts Ò 2019 Multilingual detection of hate speech against immigrants and women in Twitter at SemEval-2019 task 5: frequency analysis interpolation for hate in speech detection *Proc. 13th Int. Workshop on Semantic Evaluation* pp 460–3

[96] Pereira-Kohatsu J C, Quijano-Sánchez L, Liberatore F and Camacho-Collados M 2019 Detecting and monitoring hate speech in Twitter *Sensors* **19** 4654

[97] Alfina I, Mulia R, Fanany M I and Ekanata Y 2017 Hate speech detection in the Indonesian language: a dataset and preliminary study *2017 Int. Conf. on Advanced Computer Science and Information Systems (ICACSIS)* (Piscataway, NJ: IEEE) pp 233–8

[98] Ibrohim M O and Budi I 2019 Multi-label hate speech and abusive language detection in Indonesian Twitter *Proc. 3rd Workshop on Abusive Language Online* pp 46–57

[99] Bohra A, Vijay D, Singh V, Akhtar S S and Shrivastava M 2018 A dataset of Hindi–English code-mixed social media text for hate speech detection *Proc. 2nd Workshop on Computational Modeling of People's Opinions, Personality, and Emotions in Social Media* pp 36–41

[100] Romim N *et al* 2021 Hate speech detection in the bengali language: a dataset and its baseline evaluation *Proc. Int. Joint Conf. on Advances in Computational Intelligence* (Berlin: Springer) pp 457–68

[101] Ishmam A and Sharmin S 2019 Hateful speech detection in public Facebook pages for the Bengali language *2019 18th IEEE Int. Conf. on Machine Learning and Applications (ICMLA)* (Piscataway, NJ: IEEE) pp 555–60

[102] Mandl T, Modha S, Kumar M A and Chakravarthi B R 2020 Overview of the HASOC Track at FIRE 2020: hate speech and offensive language identification in Tamil, Malayalam, Hindi, English and German *Forum for Information Retrieval Evaluation* pp 29–32

[103] Pathak V, Joshi M, Joshi P, Mundada M and Joshi T 2021 *KBCNMUJAL@HASOC-Dravidian-CodeMix-FIRE2020: Using Machine Learning for Detection of Hate Speech and Offensive Uode-Mixed Social Media Text* (arXiv:2102.09866 [cs.CL])

[104] Saha D, Paharia N, Chakraborty D, Saha P and Mukherjee A 2021 Hate-Alert@DravidianLangTech-EACL2021: ensembling strategies for transformer-based offensive language detection *Proc. 1st Workshop on Speech and Language Technologies for Dravidian Languages* https://aclanthology.org/2021.dravidianlangtech-1.38

[105] Roy P K, Bhawal S and Subalalitha C N 2022 Hate speech and offensive language detection in Dravidian languages using deep ensemble framework *Comput. Speech. Lang.* **75** 101386

[106] Marreddy M, Oota S R, Vakada L S, Chinni V C and Mamidi R 2022 Am i a resource-poor language? Data sets, embeddings, models and analysis for four different NLP tasks in Telugu language *Transactions on Asian and Low-Resource Language Information Processing*

[107] Lee J, Lim T, Lee H, Jo B, Kim Y, Yoon H and Han S C 2022 K-MHaS: a multi-label hate speech detection dataset in Korean online news comment *Proc. 29th Int. Conf. on Computational Linguistics* pp 3530–8 https://aclanthology.org/2022.coling-1.311

[108] Jeong Y, Oh J, Ahn J, Lee J, Mon J, Park S and Oh A 2022 *KOLD: Korean Offensive Language Dataset* (arXiv:2205.11315 [cs.CL])

[109] Chen Y, Zhou Y, Zhu S and Xu H 2012 Detecting offensive language in social media to protect adolescent online safety *2012 Int. Conf. on Privacy, Security, Risk and Trust and 2012 Int. Conf. on Social Computing* (Piscataway, NJ: IEEE) pp 71–80

[110] Van Hee C, Lefever E, Verhoeven B, Mennes J, Desmet B, De Pauw G, Daelemans W and Hoste V 2015 Detection and fine-grained classification of cyberbullying events *Proc. Int. Conf. Recent Advances in Natural Language Processing* pp 672–80

[111] Zimmerman S, Kruschwitz U and Fox C 2018 Improving hate speech detection with deep learning ensembles *Int. Conf. on Language Resources and Evaluation* https://aclanthology.org/L18-1404.pdf

[112] Saha P, Mathew B, Goyal P and Mukherjee A 2018 *Hateminers: DetectingHate Speech Against Women* (arXiv:1812.06700 [cs.SI])

[113] Gao L and Huang R 2017 Detecting online hate speech using context aware models *Proc. Int. Conf. Recent Advances in Natural Language Processing, RANLP 2017 Varna, Bulgaria, September 2017* (Shumen: INCOMA Ltd) pp 260–6

[114] Zhang Z, Robinson D and Tepper J 2018 Detecting hate speech on Twitter using a convolution-GRU based deep neural network *The Semantic Web* ed A Gangemi, R Navigli, M-E Vidal, P Hitzler, R Troncy, L Hollink, A Tordai and M Alam (Cham: Springer International Publishing) pp 745–60

[115] Aluru S S, Mathew B, Saha P and Mukherjee A 2020 *Deep Learning Models for Multilingual Hate Speech Detection* (arXiv:2004.06465 [cs.SI])

[116] Caselli T, Basile V, Mitrović J and Granitzer M 2020 *HateBERT: Retraining BERT for Abusive Language Detection in English* (arXiv:2010.12472 [cs.CL])

[117] Waseem Z and Hovy D 2016 Hateful symbols or hateful people? Predictive features for hate speech detection on Twitter *Proc. NAACL Student Research Workshop* pp 88–93

[118] Davidson T, Bhattacharya D and Weber I 2019 Racial bias in hate speech and abusive language detection datasets *Proc. 3rd Workshop on Abusive Language Online* (Minneapolis, MN: Association for Computational Linguistics) pp 25–35

[119] Waseem Z, Davidson T, Warmsley D and Weber I 2017 Understanding abuse: a typology of abusive language detection subtasks *Proc. 1st Workshop on Abusive Language Online* (Vancouver, BC: Association for Computational Linguistics) pp 78–84 https://aclanthology.org/W17-3012

[120] Artetxe M and Schwenk H 2019 Massively multilingual sentence embeddings for zero-shot cross-lingual transfer and beyond *Trans. Assoc. Comput. Linguist.* **7** 597–610

[121] Agrawal S and Awekar A 2018 Deep learning for detecting cyberbullying across multiple social media platforms ed G Pasi, B Piwowarski, L Azzopardi and A Hanbury *Advances in Information Retrieval* (Cham: Springer International Publishing) pp 141–53

[122] Badjatiya P, Gupta S, Gupta M and Varma V 2017 Deep learning for hate speech detection in tweets *Proc. 26th Int. Conf. on World Wide Web Companion, WWW'17 Companion Republic and Canton of Geneva, CHE* International World Wide Web Conferences Steering Committee pp 759–60

[123] Sap M, Card D, Gabriel S, Choi Y and Smith N A 2019 The risk of racial bias in hate speech detection *Proc. 57th Annual Meeting of the Association for Computational Linguistics* (Minneapolis, MN: Association for Computational Linguistics) pp 1668–78

[124] Shah D J, Wang S, Fang H, Ma H and Zettlemoyer L 2021 *Reducing Target Group Bias in Hate Speech Detectors* (arXiv:2112.03858 [cs.CL])

[125] Pavlopoulos J, Sorensen J, Laugier L and Androutsopoulos I 2021 SemEval-2021 task 5: toxic spans detection *Proc. 15th Int. Workshop on Semantic Evaluation (SemEval-2021)* pp 59–69

[126] Chandrasekharan E, Pavalanathan U, Srinivasan A, Glynn A, Eisenstein J and Gilbert E 2017 You can't stay here: the efficacy of Reddit's 2015 ban examined through hate speech *Proc. ACM on Human–Computer Interaction* vol 1 CSCW, pp 1–22

[127] Ali S, Saeed M H, Aldreabi E, Blackburn J, De Cristofaro E, Zannettou S and Stringhini G 2021 Understanding the effect of deplatforming on social networks *13th ACM Web Science Conf. 2021* pp 187–95

[128] Mathew B, Saha P, Tharad H, Rajgaria S, Singhania P, Maity S K, Goyal P and Mukherjee A 2019 Thou shalt not hate: countering online hate speech *Proc. Int. AAAI Conf. on Web and Social Media* **vol 13** pp 369–80

[129] Mathew B, Kumar N, Goyal P and Mukherjee A 2020 Interaction dynamics between hate and counter users on Twitter *Proc. 7th ACM IKDD CoDS and 25th COMAD* pp 116–24

[130] Chung Y-L, Kuzmenko E, Tekiroglu S S and Guerini M 2019 CONAN—COunter NArratives through Nichesourcing: A Multilingual Dataset of Responses to Fight Online Hate Speech (arXiv:1910.03270 [cs.CL])

[131] Tekiroğlu S S, Chung Y-L and Guerini M 2020 Generating counter narratives against online hate speech: data and strategies *Proc. 58th Annual Meeting of the Association for Computational Linguistics* (Minneapolis, MN: Association for Computational Linguistics) pp 1177–90

[132] Zhu W and Bhat S 2021 *Generate, Prune, Select: A Pipeline for Counterspeech Generation Against Online Hate Speech* (arXiv:2106.01625 [cs.CL])

[133] Bowman S R, Vilnis L, Vinyals O, Dai A M, Jozefowicz R and Bengio S 2015 *Generating Sentences from a Continuous Space* (arXiv:1511.06349 [cs.LG])

[134] Warstadt A and Bowman S R 2019 *Linguistic Analysis of Pretrained Sentence Encoders with Acceptability Judgments* (arXiv:1901.03438 [cs.CL])

[135] Henderson M, Vulić I, Gerz D, Casanueva I, Budzianowski P, Coope S, Spithourakis G, Wen T-H, Mrkšić N and Su P-H 2019 *Training Neural Response Selection for Uask-Oriented Dialogue Systems* (arXiv:1906.01543 [cs.CL])

[136] Saha P, Singh K, Kumar A, Mathew B and Mukherjee A 2022 CounterGeDi: a controllable approach to generate polite, detoxified and emotional counterspeech; L De Raedt *Proc. 31st Int. Joint Conf. on Artificial Intelligence, IJCAI-22* pp 5157–63

[137] Lacy S, Fico F and Simon T F 1991 Fairness and balance in the prestige press *Journal. Q* **68** 363–70

[138] Riffe D, Lacy S and Fico F 2014 *Analyzing Media Messages: Using Quantitative Content Analysis in Research* (London: Routledge)

[139] Bachmann P, Hunziker S and Rüedy T 2019 Selling their souls to the advertisers? How native advertising degrades the quality of prestige media outlets *J. Media Bus. Stud.* **16** 95–109

[140] Park S, Kang S, Chung S and Song J 2009 NewsCube: delivering multiple aspects of news to mitigate media bias *Proc. SIGCHI Conf. on Human Factors in Computing Systems* pp 443–52

[141] Ribeiro F N, Henrique L, Benevenuto F, Chakraborty A, Kulshrestha J, Babaei J and Gummadi K P 2018 Media bias monitor: quantifying biases of social media news outlets at large-scale *12th Int. AAAI Conf. on Web and Social Media*

[142] Chakraborty A, Sarkar R, Mrigen A and Ganguly N 2017 Tabloids in the era of social media? Understanding the production and consumption of clickbaits in Twitter *Proc. ACM on Human–Computer Interaction* vol 1 (New York: Association for Computing Machinery) pp 30:1–30:21

[143] Manning C D 2008 *Introduction to Information Retrieval* (Oxford: Syngress Publishing)

IOP Publishing

AI and Ethics
A computational perspective
Animesh Mukherjee

Chapter 5

Interpretable versus explainable models

5.1 Chapter foreword

This chapter will introduce the concepts of interpretability and explainability in AI. We shall start by outlining an intuition of each of these buzzwords and discuss their primary differences. Next, we shall attempt to operationalize the definition of each of these and demonstrate how machine learning models could be made interpretable or explainable.

5.2 What is interpretability?

It is in human nature to routinely create interpretations about their surroundings. In AI, such models can be built based on these interpretations and further tested to verify if they are accurate representations of the world. Precisely, interpretability means that it should be possible to associate a cause with an effect. The thought of interpretability always reminds me of a classic example from Agatha Christie's detective fiction The Pale Horse. It took Mark Easterbrook, the central character, to come up with an interpretable mental model to connect the cause-and-effect resulting in the series of murders narrated in the story. While the adversaries were eager to establish that death could be caused by a strong desire and from a distance using powers of witchcraft, Mark finally came up with a curious observation that all victims suffered a loss of hair or their hair could be pulled out easily from their scalp close to their death (an unusually difficult task in general). He was finally able to recognize that this was a recurring symptom of thallium poisoning and not a case of 'satanic assassination', thus solving the mystery. Such is usually the power of having an interpretable model.

In the case of AI models, they are considered to be interpretable if their predictions are easily understandable and therefore trusted by humans. For instance, if there is a model that predicts the mortality risk based on patient records, then that model is said to be interpretable if one can easily infer why the model predicted a particular mortality risk for a particular patient. We illustrate this in figure 5.1 where

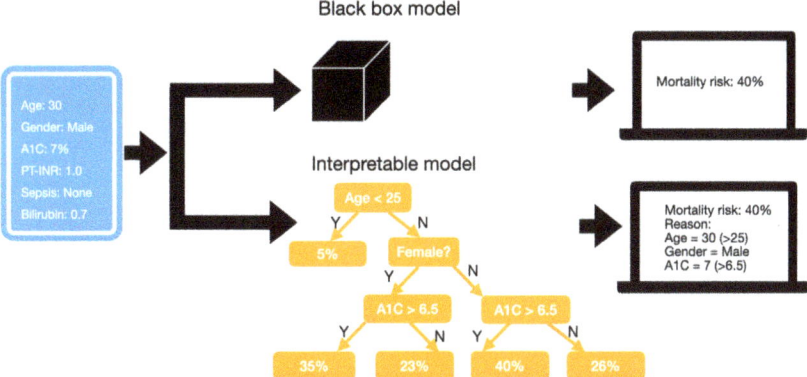

Figure 5.1. The difference between a black box and an interpretable model.

the black box model predicts the mortality risk to be 40% for the input patient data without giving any reason while the interpretable decision tree based model makes the same prediction but with precise reasons.

5.3 What is explainability?

Neural models for ML are often termed black boxes since they allow for a set of empty parameters aka nodes that are updated by the algorithm usually through backpropagation. Explainability refers to the extent to which a particular node is responsible for the performance of a model. A pertinent question one can ask here is why is explainability at all important given the skyrocketing performance accuracy of the modern deep learning models? One of the prime motivating examples comes from the task of image classification where the model has as input a sequence of pixels and a label associated with each image. Thus if the input image is of a cat, the model should be able to predict the label 'cat' with very high confidence. However, there are numerous instances where models have resulted in highly undesirable results. In 2015, Google Photos tagged two African-Americans as 'gorillas' based on its black box facial recognition software[1]. Three years later in 2018, an investigation[2] carried out by *Wired* showed that the company had blocked its image recognition software from identifying gorillas altogether. Over tens of thousands of pictures of primates were uploaded by the *Wired* team upon which they found that while baboons, gibbons, and marmosets were identified correctly, gorillas and chimpanzees were not. Thus it seemed that the company resorted to an ad hoc solution rather than deep diving and fixing the roots of the problem. As recently as in 2020, AlgorithmWatch[3] demonstrated how the computer vision service Google Vision Cloud labeled the image of a thermometer held by a dark-skinned person as a 'gun'

---

[1] https://tinyurl.com/2zmtm8hr
[2] https://tinyurl.com/4mkjtj6u
[3] https://algorithmwatch.org/en/

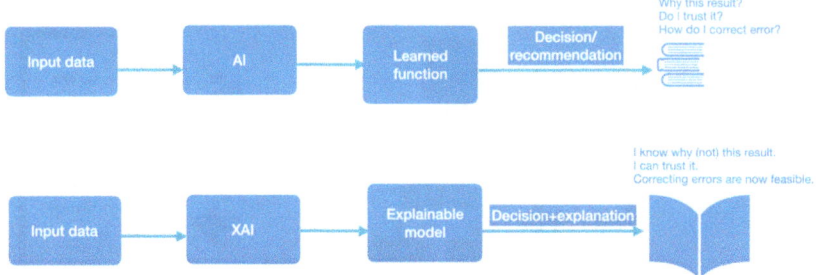

Figure 5.2. The difference between a regular and an explainable AI model.

while a similar image held by a light-skinned person as an 'electronic device'[4]. As a subsequent step the team observed that if one 'lightens' the dark-skin hand in the former image with a salmon-colored mask, the algorithm switches the label from 'gun' to 'monocular'. In [1] the authors noted how biases in facial recognition software from Amazon and Microsoft become pronounced if the images are adversarially perturbed. Thanks to journalists, researchers, and activists, big techs are constantly kept in check from committing such errors that result in biased decisions based on sex, race, etc. However, a principled computational approach needs to be in place that can explain the predictions of an ML model. This has led to the inception of a new avenue of research–XAI (explainable AI)—that allows humans to understand, trust, and suitably manage their AI assistants (see figure 5.2).

5.4 Operationalizing interpretability

There are various traditional approaches to operationalize the concept of interpretability. Here, we shall discuss some of the most popular techniques.

5.4.1 Partial dependence plot

The partial dependence plot (aka PDP) [2] was invented around 20 years ago and attempts to capture the marginal effect of at most one or two features on the predictions made by a machine learning model $\hat{f}$. To achieve this the feature vector **x** is divided into two parts—$\mathbf{x}_S$ and $\mathbf{x}_C$. Here S is the set of features (usually one or two) for which the partial dependence function $\hat{f}_S$ is to be computed and C are the other features. The central idea is to marginalize the output of the machine learning model over the distribution of the feature set C, so that the function $\hat{f}_S$ unfolds the relationship between the feature set S and the model prediction. The partial dependence function is usually computed by taking averages over the training data and can be expressed as follows

$$\hat{f}_S(x_S) = \sum_{k=1}^{n} \hat{f}(x_S, x_C^{(k)}) \tag{5.1}$$

---

[4] https://algorithmwatch.org/en/google-vision-racism/

In the above equation, $x_C^{(k)}$ are the feature values from the dataset which we are not interested in and n is the number of instances in the dataset. PD plots consider the feature(s) of interest for a specified range. For each value of the feature, the model is evaluated for all the observations of the other inputs, and the output is then averaged. Note that PDP assumes that the features in C and S are uncorrelated; the violation of this might result in having very unlikely data points included in the calculation of the average.

Feature importance: based on the concept of PDP, it is possible to compute the importance of the different features. This was developed by the authors in [3] based on the observation that a feature is not important if the corresponding PDP is flat; the higher the PDP varies, the more important the feature is. If the feature values are numerical, this can be expressed as the deviation of each feature value from the average as follows

$$I(x_S) = \sqrt{\frac{1}{K-1}\sum_{i=1}^{K}\left(\hat{f}_S(x_S^{(i)}) - \frac{1}{K}\sum_{i=1}^{K}\hat{f}_S(x_S^{(i)})\right)^2} \qquad (5.2)$$

In the above equation, $x_S^{(i)}$ denotes the K unique feature values of x_S.

Examples: PDPs have been widely used in the literature to establish the dependence of the model output on a particular input of interest. For instance, one can identify whether the chances of a flu infection increase linearly with fever or whether the rise of body temperature reduces the probability of flu. It can also indicate more nuanced items like whether the price of a car is linearly dependent on the horsepower or if the relationship is more complex such as a step function or even curvilinear.

5.4.2 Functional decomposition

Any supervised machine learning model is a function that takes as input a high-dimensional feature vector and produces a classification output or a prediction score. Functional decomposition attempts to deconstruct this function into a sum of the effects of the individual features and the effects of their interactions. For instance, if we have only two input features x_i and x_j then the learning function $\hat{f}(x_i, x_j)$ can be decomposed as follows.

$$\hat{f}(x_i, x_j) = \hat{f}_0 + \hat{f}_i(x_i) + \hat{f}_j(x_j) + \hat{f}_{i,j}(x_i, x_j) \qquad (5.3)$$

Here $\hat{f}_i$ and $\hat{f}_j$ are the main effects, $\hat{f}_{i,j}$ is the interaction and $\hat{f}_0$ is the intercept. Generalizing for a d-dimensional feature vector with $\hat{f}: \mathbb{R}^d \to \mathbb{R}$ we have

$$\begin{aligned}\hat{f}(x) =\ & \hat{f}_0 + \hat{f}_1(x_1) + \cdots + \hat{f}_d(x_d) \\ & + \hat{f}_{1,2}(x_1, x_2) + \cdots + \hat{f}_{1,d}(x_1, x_d) + \cdots + \hat{f}_{d-1,d}(x_{d-1}, x_d) \\ & + \cdots \\ & + \hat{f}_{1,\ldots,d}(x_1,\ldots,x_d)\end{aligned} \qquad (5.4)$$

We can rewrite the above equation in a more concise form as follows. Let $S \subseteq \{1,\ldots,d\}$ be the subsets of feature combinations, i.e., $S = \phi$ corresponds to the intercept, $|S| = 1$ corresponds to the main effect and $|S| > 1$ correspond to all the interactions. Based on this the decomposition formula takes the following form

$$\hat{f}(x) = \sum_{S \subseteq \{1,\ldots,d\}} \hat{f}_S(x_S) \tag{5.5}$$

The hardship of functional decomposition is that one has to compute an exponential number of components, i.e., for a d-dimensional feature space, the number of subsets in the above formula would be 2^d. In order to circumvent this problem one of the popular ways is to resort to statistical regression that we discuss next.

5.4.3 Statistical regression

We use statistical regression to restrict the number of components so that not all 2^d of them need to be computed as per functional decomposition. Linear regression is the simplest among these and can be expressed as follows

$$\hat{f}(x) = \beta_0 + \beta_1 x_1 + \cdots \beta_d x_d \tag{5.6}$$

The above equation is very similar to that of functional decomposition, however, with two important changes. First, only the main effects and the intercept are retained while the interaction effects are discarded. Second, the main effects are assumed to be linear of the form $\hat{f}_j(x_j) = \beta_j x_j$. It is possible to relax the second assumption and include other functions by using generalized additive model [4].

5.4.4 Individual conditional expectation

Individual conditional expectation (ICE) [5] plots demonstrate how the prediction for an instance changes when a feature changes. There is one line for each instance in the plot. The ICE plot is equivalent to a PDP plot for individual data instances. In other words, a PDP plot is the average of all the lines in an ICE plot. Formally, for every instance $\{(x_S^{(i)}, x_C^{(i)})\}_{i=1}^{n}$ the function $\hat{f}_S^{(i)}$ is plotted against $x_S^{(i)}$ keeping $x_C^{(i)}$ fixed. If features are correlated among themselves, ICE gives much better insights into the workings of the models than PDP.

Centered ICE plot: often it is difficult to compare the individual ICE plots as they start at different prediction values. One way to tackle this is to center the plots on a single point in the feature and only show the difference between this point and the prediction. Such a plot is called a centered ICE plot (c-ICE). One of the simplest options could be to anchor the curves to the lower end of the feature. Hence, the new curves take the following form

$$\hat{f}_{cent}^{(i)} = \hat{f}^{(i)} - \mathbf{1}\hat{f}(x^a, x_C^{(i)}) \tag{5.7}$$

In the above equation, $\hat{f}$ is the learnt function, x^a is the anchor point and $\mathbf{1}$ is the vector of all 1's.

5.5 Operationalizing explainability

Recent studies in the AI/ML literature have invested a great deal of effort to develop techniques that can explain the workings of the black box models. In the following, we shall discuss some of the most popular techniques.

5.5.1 Local surrogate

Local surrogate techniques are developed to explain individual predictions of black box machine learning models. Local interpretable model-agnostic explanations (LIME) [6] is one of the most popular local surrogate models in the literature. Typically a surrogate model, as the name suggests, is trained to approximately mimic the predictions of the back-end black box model. The surrogate may be either a global one to explain overall predictions or a local one to explain individual predictions. As the name suggests, LIME develops a local surrogate technique.

The idea of LIME is very simple and intuitive. It perturbs the original training data and probes the black box model to obtain the predictions for each of the new data points resulting in a new perturbed dataset. Next it trains an interpretable model (e.g., decision tree) on this perturbed dataset where the learning is weighted by the proximity of the sampled data points from the actual point of interest for which an explanation is needed. The hypothesis is that the learned interpretable model should be a good approximation of the actual black box model.

The equation below represents the local surrogate model with interpretability constraints

$$\text{expl}(x) = \arg\min_{e \in E} L(f, e, \pi_x) + \Omega(e) \tag{5.8}$$

For a given instance x, the explanation corresponds to a model e (say linear regression) that minimizes a loss function (say mean squared error), which computes the proximity of the explanation to the prediction of the black box model f (say a neural network) keeping the overall complexity of the model, $\Omega(e)$, low. π_x corresponds to the size of the neighborhood around the instance x that we consider to generate the explanation while E is the family of explanations (say all possible linear regression models). Algorithm 2 below summarizes how the local surrogate model is trained in the LIME technique.

Algorithm 2: Algorithm to train the local surrogate model.

Require Instance x for which an explanation is sought for its black box prediction
    Perturb the dataset to obtain a new set of instances
    Compute black box predictions for these n new instances
    Assign weights to new instances in accordance with their proximity to x
    Train the weighted model on the new data points through the learning function noted in equation (5.8)
    Return $\text{expl}(x)$

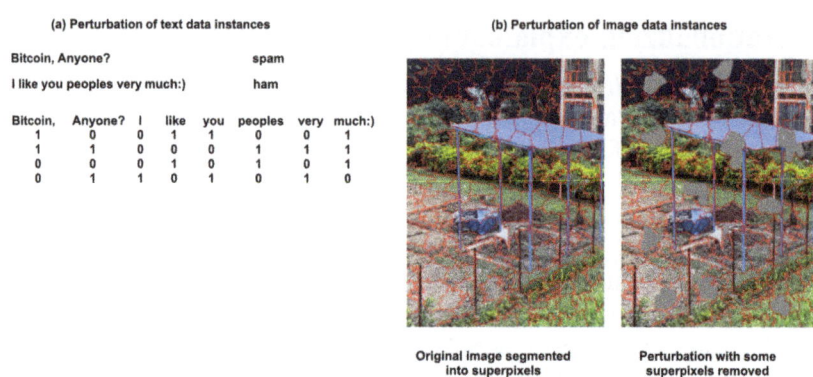

Figure 5.3. Generating variants of (a) text and (b) image data.

Generating variants for text data: for textual data, variants can be obtained by removing words randomly from the original text. The presence (absence) of each word is represented by a 1(0) feature in the dataset. If a particular word is retained after perturbation, the corresponding feature is 1 else it is 0. Figure 5.3(a) shows an example data meant for spam/ham message detection; some of the instances after perturbation are also shown.

Generating variants for image data: for image data, perturbing individual pixels does not make much sense since individual classes are constituted by more than one pixel. Thus, changing a random set of pixels will possibly not impact the class probabilities. Hence, in the case of images, the perturbations are done at the level of 'superpixels'. Superpixels correspond to a set of interconnected pixels with similar colors. These pixels can be together turned off by changing all of them with a user-defined color, e.g., gray. Perturbations are generated by switching one or more superpixels as a whole. Figure 5.3(b) shows an original image segmented into superpixels as well as a random perturbation where some of the superpixels have been converted into gray.

5.5.2 Anchors

The main idea here is to find a decision rule that 'anchors' the prediction of a black box model sufficiently. A rule is said to anchor a prediction if changes in other feature values do not affect the prediction outcome. This is an extension of the LIME technique and was developed by the same group of authors [7]. The explanation in the LIME technique is usually not faithful since it learns only a decision boundary from the perturbed feature space. The anchor-based approach works on the same perturbed feature space but is clearly able to delineate its boundaries by stating the instances for which the predictions are valid. This makes the predictions more intuitive and easy to understand. An anchor can be expressed by the following equation

$$\mathbb{E}_{\mathcal{D}_x(z|A)}[1_{\hat{f}(x)=\hat{f}(z)}] \geq \tau, \ A(x) = 1 \tag{5.9}$$

The idea behind the equation is as follows. One needs to find an explanation for the instance x and has the task to find an anchor A for this purpose. The anchor A should be applicable to x and also to at least a fraction τ of the neighbors of x that have the same predicted class as x. The precision of the rule A is obtained by evaluating the neighbors/perturbations, i.e., $\mathcal{D}_x(z|A)$ using the ML model $\hat{f}$ (the equivalence check of the class labels of x and one of its neighbor z is expressed by the indicator function $1_{\hat{f}(x)=\hat{f}(z)}$).

It is, however, difficult to evaluate $1_{\hat{f}(x)=\hat{f}(z)}$ for all $z \in \mathcal{D}_x(\cdot|A)$ given that the dataset is very large. Equation (5.9) is therefore modified in the following way

$$P(\text{prec}(A) \geq \tau) \geq 1 - \delta \quad \text{with} \quad \text{prec}(A) = \mathbb{E}_{\mathcal{D}_x(z|A)}[1_{\hat{f}(x)=\hat{f}(z)}] \qquad (5.10)$$

Equation (5.10) is a probabilistic version of equation (5.9) that introduces the parameter $0 \leq \delta \leq 1$ so that the samples are drawn until statistical confidence on the precision is obtained. The two equations are connected through the notion of *coverage* which indicates the probability of the application of the anchor to explain the neighbors of x. Formally, coverage is defined as follows

$$\text{cov}(A) = \mathbb{E}_{\mathcal{D}_{(z)}}[A(z)] \qquad (5.11)$$

Thus the final definition of the anchor in terms of the coverage can be expressed as follows

$$\max_{A \text{ s.t. } P(\text{prec}(A) \geq \tau) \geq 1-\delta} \text{cov}(A) \qquad (5.12)$$

Equation (5.12) maximizes the coverage over all eligible rules that satisfy the precision threshold τ under the probabilistic definition. The equation attempts to find a trade-off between the precision of the anchor and its coverage. The candidate rules are compared to check which of them best explains the perturbations around x by calling the prediction model multiple times. To make this process faster, the authors effectively used the multi-armed bandit paradigm and the results over the multiple rounds are assembled using a beam search technique.

5.5.3 Shapley values

The explanation of a prediction can be obtained by imagining each feature of the instance as a 'player' in a game and the prediction as the 'payout'. The idea is to compute the performance of a team in the game with and without a particular player P. The impact of the player P can be measured as follows—Impact of P = Performance of the team with P—Performance of the team without P. The same idea can be extended to each individual player to compute their respective impacts. Additive Shapley values popular in the coalition game theory literature [8] present an elegant way to compute this impact. Shapley value corresponds to the average marginal contribution of a particular feature value across all possible feature combinations. Extending on the previous example let us assume a three-player team—P_1, P_2, and P_3. We wish to compute the contribution of player P_1

when it is added to the coalition of the players P_2 and P_3. The experimental setup requires that all trials with and without each player are available for the estimation of importance. Given these, we can now have the following configurations:

Step 1: P_2 and P_3 participate in a trial in absence of P_1. This can be approximated as the average score/payout in all matches where P_2 and P_3 participated but P_1 did not. Let us say the average score is 105.

Step 2: P_2 and P_3 participate in a trial along with P_1. Once again this can be approximated as the average score/payout in all matches where P_1, P_2, and P_3 participated. Let us say the average score is 135.

The contribution of P_1 is calculated as the difference between the two scores, i.e., $135 - 105 = 30$.

The Shapley value for a feature i is defined as follows

$$\phi_i = \sum_{S \subseteq P \cup \{i\}} \frac{|S|!(|P| - |S| - 1)!}{|P|!} \left(\hat{f}(S \cup \{i\}) - \hat{f}(S) \right) \tag{5.13}$$

In this equation, S refers to the subset of features that does not include the feature for which we compute ϕ_i. $S \subseteq P \cup \{i\}$ corresponds to all subsets S of the full feature set P excluding the feature i. $S \cup \{i\}$ is the subset of features in S plus the feature i. $\hat{f}(S)$ is the prediction for the feature values in set S that are marginalized over features that are not part of set S. In practice, the approximate Shapley value for a feature i is computed using a Monte-Carlo approach through the following equation

$$\hat{\phi}_i = \frac{1}{N} \sum_{n=1}^{N} \left(\hat{f}(x_{+i}^n) - \hat{f}(x_{-i}^n) \right) \tag{5.14}$$

Here $\hat{f}(x_{+i}^n)$ is the prediction for the instance x but with a random number of its feature values changed with those features of another random instance z however keeping the feature at index i the same. $\hat{f}(x_{-i}^n)$ is very similar but here the feature at index i is also changed with the corresponding feature of the random instance z. Algorithm 3 presents a method for the approximate estimation of the Shapley value of a single feature.

Overall the algorithm works as follows. First, the instance x, the feature index i, the number of iterations N, and the data matrix D are taken as input. The black box predictor $\hat{f}$ is also known. In every iteration, another instance z is chosen randomly as well as a random order of the features is generated. From the two instances, x and z two new instances are generated. The instance x_{+i} is the instance of interest with the exception that all values in the order after feature i are replaced by those from the sample z. The instance x_{-i} is the same as x_{+i} with the exception that the value at the feature index i is also taken from the sample z. Next the differences with the black box predictor ϕ_n^i are computed and averaged over all the iterations. This procedure is repeated for every feature index separately to obtain the Shapley values for these features.

5.5.4 SHAP

SHAP (SHapley Additive exPlanations) [9] attempts to generate explanations for individual predictions. The basic idea is to compute the contribution of each feature to the prediction of an instance x thus explaining the prediction. SHAP is based on the same principle of computing Shapley values where the explanation is computed as follows

$$g(c') = \phi_0 + \sum_{i=1}^{N} \phi_i c_i' \qquad (5.15)$$

Algorithm 3: Approximate Shapley value estimation.

Input: Number of iterations N
Input: Instance of interest x
Input: Feature index i
Input: Machine learning model $\hat{f}$
Data: Data matrix D
Result: $\phi_i(x)$
for $n = 1 \ldots N$ **do**
  Pick an instance z randomly form D;
  Decide a permutation o of the feature values;
  Instance x ordered as: $x_o = (x_{(1)}, \ldots, x_{(i)}, \ldots, x_{(p)})$;
  Instance z ordered as: $z_o = (z_{(1)}, \ldots, z_{(i)}, \ldots, z_{(p)})$;
  Two new instances are constructed:
  with i: $x_{+i} = (x_{(1)}, \ldots, x_{(i-1)}, x_{(i)}, z_{(i+1)}, \ldots, z_{(p)})$;
  without i: $x_{-i} = (x_{(1)}, \ldots, x_{(i-1)}, z_{(i)}, z_{(i+1)}, \ldots, z_{(p)})$;
  Marginal contribution: $\phi_i^n = \hat{f}(x_{+i}) - \hat{f}(x_{-i})$
end
Approximate Shapley value: $\phi_i(x) = \frac{1}{N} \sum_{n=1}^{N} \phi_i^n$;

---

where g is the explanation model and $c' \in \{0, 1\}^N$ is the simplified set of features (or a *coalition*), N is the maximum size of the coalition and $\phi_i \in \mathbb{R}$ is the Shapley value for the feature i. c' is a vector array where an entry of 1 corresponds to the presence of a feature and 0 to its absence (similar to the idea of Shapley values). For x which is the instance of interest, x' is a vector of all 1s that is all the features are present. In this case, the above equation (5.15) reduces to the following

$$g(x') = \phi_0 + \sum_{i=1}^{N} \phi_i \qquad (5.16)$$

In order to compute SHAP the authors in [9] proposed the algorithm KERNELSHAP that estimates for an instance x the contribution of each feature to the prediction. Algorithm 4 describes the KERNELSHAP method. A random coalition can be created by random coin tosses repeatedly until a requisite number of 1s and 0s are obtained. For instance, if we have a sequence $\langle 0, 1, 0, 0, 1 \rangle$ then it means that there is a coalition in the second and the fifth features. The sampled coalitions $k \in \{1,...,K\}$ become the dataset for the regression model. One needs to have a function to map the coalitions of feature values to valid data instances. In order to do this, the authors proposed the function $h_x(c') = c$ where $h_x: \{0, 1\}^N \to \mathbb{R}^p$ which maps 1s to the corresponding feature values from the instance x which needs to be explained. The 0s are mapped to the feature values of another randomly sampled data as was done for the Shapley value computation.

Algorithm 4: The KERNELSHAP algorithm.

Sample $c_k' \in \{0, 1\}^M$, $k \in \{1,...,K\}$; $\triangleright$ 1(0) = feature present(absent)
Convert each c_k' to the original feature space;
Get predictions for the converted c_k' by applying the model $\hat{f}: \hat{f}(h_x(c_k'))$;
The weights of c_k's are computed using the SHAP kernel;
The weighted linear model g is fitted;
return ϕ_k; $\triangleright$ the coefficients from the linear model.

The linear model g is fitted by optimizing the following loss function

$$L(\hat{f}, g, \pi_x) = \sum_{c' \in Z} [\hat{f}(h_x(c')) - g(c')]^2 \pi_x(c') \tag{5.17}$$

where $g(c')$ is defined as in equation (5.15).

5.5.5 Counterfactual explanations

Counterfactual analysis is based on the idea of constructing *what-if?* scenarios to evaluate the outcomes that did not occur in reality but could have occurred under different conditions. According to the Merriam-Webster dictionary, the word counterfactual refers to 'contrary to facts'. Counterfactual reasoning allows for the identification of cause-and-effect relationships. Counterfactual reasoning requires imagining a hypothetical world that contradicts the observed outcomes. An example could be the following counterfactual reasoning that one could make after a car crash—'if the driver was not speeding, my car would not have been wrecked'. In this case, the hypothetical world is where 'the driver had not sped up the car'.

In explainable machine learning models, counterfactuals are simulated in the following way—one observes that if a small change is made in a feature value corresponding to an instance how does the prediction outcome change? The prediction outcome should change meaningfully; for instance, causing the predicted class label to be flipped.

Generation of counterfactual explanations by Watcher et al: Watcher et al [10] proposed a simple method to generate counterfactual explanations for any standard machine learning algorithm. Recall that any machine learning algorithm is trained by computing an optimal set of weights w that minimizes a loss function L over the training data. This takes the following form

$$\arg \min_{w} L(\hat{f}_w(x), y) + \rho(w) \tag{5.18}$$

In this equation y is the target label for the data point x and $\rho(\cdot)$ is a regularizer over the weights. Our objective is to compute a counterfactual x' which is as close to the original instance x as possible under the requirement that $\hat{f}_w(x')$ is equal to a new target y'. Typically x' is searched by holding w fixed and minimizing the related objective below (we will get rid of w since it is held fixed)

$$L(x, x', y', \lambda) = \lambda \cdot (\hat{f}(x') - y')^2 + d(x, x') \tag{5.19}$$

The desired outcome y' needs to be predefined by the user. $d(\cdot, \cdot)$ is a function that measures the distance between the instance of interest x and the counterfactual x'. Thus the loss function has two components: (i) how far the predicted outcome of the counterfactual is from the predefined outcome y' and (ii) how far the counterfactual x' is from the instance of interest x. The parameter λ strikes a balance between the prediction distance (first term) and the feature distance (second term). The function d is defined as follows

$$d(x, x') = \sum_{i=1}^{|F|} \frac{|x_i - x'_i|}{\mathrm{MAD}_i} \tag{5.20}$$

In this equation F is the set of all features, i is a specific feature and MAD_i is the median absolute deviation of feature i.

In practice, the parameter λ is maximized by iteratively increasing its value and solving for x' until a sufficiently close solution is obtained. In order to avoid getting stuck to a local minima, the authors randomly initialized x' and selected that counterfactual which minimized the following equation

$$\mathrm{argmin}_{x'} \max_{\lambda} L(x, x', y', \lambda). \tag{5.21}$$

Instead of directly tuning λ, the authors proposed to tune a tolerance parameter ϵ such that $|\hat{f}(x') - y'| \leqslant \epsilon$. The optimization can be performed using a standard ADAM optimizer.

Generation of counterfactual explanations by Dandl et al: The authors in this paper [11] simultaneously optimize four objective functions together as follows

$$L(x, x', y', X^{\mathrm{obs}}) = \left(o_a(\hat{f}(x'), y'), o_b(x, x'), o_c(x, x'), o_d(x', X^{\mathrm{obs}})\right) \tag{5.22}$$

The first objective o_a determines that the prediction outcome for the counterfactual x' should be as close as the predefined output y'. o_a is defined based on L_1 norm as follows

$$o_a(\hat{f}(x'), y') = \begin{cases} 0 & \text{if } \hat{f}(x') \in Y' \\ \inf_{y' \in Y'} |\hat{f}(x') - y'| & \text{otherwise} \end{cases} \quad (5.23)$$

The set $Y' \subset \mathbb{R}$ is the set of desirable outcomes, i.e., admits the desired class probability range.

The second objective ensures the closeness of the counterfactual x' and the instance of interest x. This is defined using the Gower distance [12] suitable for handling mixed features

$$o_b(x, x') = \frac{1}{|F|} \sum_{i=1}^{|F|} \delta_G(x_i, x'_i) \quad (5.24)$$

Here again F is the set of all features. δ_G takes up different forms based on the type of features as follows

$$\delta_G(x_i, x'_i) = \begin{cases} \frac{1}{\hat{R}_i}|x_i - x'_i| & \text{if } x^i \text{ numerical} \\ \mathbb{I}_{x_i \neq x'_i} & \text{if } x_i \text{ categorical} \end{cases} \quad (5.25)$$

$\hat{R}_i$ is the range of the feature value i for a give dataset and is used to rescale δ_G between 0 and 1.

Since the Gower distance does not have control on the number of features being changed, this is done by optimizing o_c defined as follows

$$o_c(x, x') = ||x - x'||_0 = \sum_{i=1}^{|F|} \mathbb{I}_{x'_i \neq x_i} \quad (5.26)$$

The minimization of o_c ensures sparse feature changes.

Finally, o_d attempts to ensure that the counterfactuals are composed of likely features or feature combinations. This is done by minimizing the Gower distance between x' and the nearest observed data point $x^{[1]} \in X^{\text{obs}}$ that approximates the likelihood of x' originating from the same feature distribution. o_d is defined as follows.

$$o_d(x', \mathbf{X}^{\text{obs}}) = \frac{1}{|F|} \sum_{i=1}^{|F|} \delta_G(x'_i, x_i^{[1]}) \quad (5.27)$$

The authors solved the above optimization using the popular non-dominated sorting genetic algorithm (NSGA-II) [13] inspired by Darwin's 'survival of the fittest' idea. The fitness of a conterfactual x' is denoted by its vector of objective values (o_a, o_b, o_c, o_d). The lower the values of these objectives for a counterfactual, the fitter it is deemed to be.

Practice problems

Q1. What is explainable AI? What is interpretable AI? How do you think are they are different?

Q2. State some use case of explainable and interpretable AI use cases.

Q3. Do you think a company needs explainable AI solutions during only reviews and audits? Justify.

Q4. Is explainable AI required in algorithmic cancer diagnosis especially in the context of medical imaging?

Q5. How can explainable AI be used for fake news detection?

Q6. How do you think explainable AI is related to trust?

Q7. Autonomous vehicles are likely to be widespread in the future. Do you believe explainable AI would have a role in the design of such autonomous vehicles? Justify.

References

[1] Jaiswal S, Duggirala K, Dash A and Mukherjee A 2022 Two-face: adversarial audit of commercial face recognition systems *Proc. Int. AAAI Conf. on Web and Social Media* vol 16 pp 381–92

[2] Friedman J H 2001 Greedy function approximation: a gradient boosting machine *Ann. Stat.* **29** 1189–232

[3] Greenwell B M, Boehmke B C and McCarthy A J 2018 *A Simple and Effective Model-Based Variable Importance Measure* arXiv:1805.04755 [stat.ML]

[4] Hastie T J 2017 Generalized additive models *Statistical Models in* S ed J M Chambers and T J Hastie (London: Routledge) pp 249–307

[5] Goldstein A, Kapelner A, Bleich J and Pitkin E 2015 Peeking inside the black box: visualizing statistical learning with plots of individual conditional expectation *J. Comput. Graph. Stat.* **24** 44–65

[6] Ribeiro M T, Singh S and Guestrin C 2016 'Why should I trust you?' Explaining the predictions of any classifier *Proc. 22nd ACM SIGKDD Int. Conf. on Knowledge Discovery and Data Mining* pp 1135–44

[7] Ribeiro M T, Singh S and Guestrin C 2018 Anchors: high-precision model-agnostic explanations *Proc. AAAI Conf. on Artificial Intelligence* vol 32

[8] Shapley L S 1997 A value for *n*-person games *Classics in Game Theory* p 69

[9] Lundberg S M and Lee S-I 2017 A unified approach to interpreting model predictions *Adv. Neural Inf. Process. Syst.* **30**

[10] Wachter S, Mittelstadt B and Russell C 2017 Counterfactual explanations without opening the black box: automated decisions and the GDPR *Harv. J. Law Technol.* **31** 841 https://papers.ssrn.com/sol3/papers.cfm?abstract_id=3063289#

[11] Dandl S, Molnar C, Binder M and Bischl B 2020 Multi-objective counterfactual explanations *Int. Conf. on Parallel Problem Solving from Nature* (Berlin: Springer) pp 448–69
[12] Gower J C 1971 A general coefficient of similarity and some of its properties *Biometrics* **27** 857–71
[13] Deb K, Pratap A and Agarwal S 2002 A fast and elitist multiobjective genetic algorithm: NSGA-II *IEEE Trans. Evol. Comput.* **6** 182–97

Chapter 6

Newly emerging paradigms

6.1 Chapter foreword

In this last chapter, we shall briefly cover some of the newly emerging paradigms. In particular, we shall discuss how law enforcement needs to change while using AI technology ethically. Next, we shall discuss how data protection, user privacy, etc, are interlinked with the design of ethical algorithms. Subsequently, we shall highlight the idea of designing *morally responsible machines*. Finally, we shall discuss how *singularity* (if any) would handshake with ethical practices in the future.

6.2 Ethical use of AI for law enforcement

AI technology is being extensively used in the law industry because of its large range of promising applications. The range of applications spans fraud detection, traffic accident monitoring, child pornography tracking, the identification of anomalies in public spaces, etc. These applications have become popular since they are less time-consuming, more cost-effective, and less vulnerable to human error and fatigue [1]. Such burgeoning use of AI in law enforcement can be attributed to their ability to identify patterns of crime and, thereby, forecast, prevent, and manage crime. The idea of crime management using AI technology is known as *predictive policing* [2, 3]. However, the use of predictive policing in the law industry has been subject to controversy, as we discuss next, because of the associated ethical and juridical concerns [4, 5].

6.2.1 Ethical concerns in predictive policing

Two important ethical questions arise in the deployment of predictive policing. First, what is the intended purpose of the use of policing, and who are the targets of policing? For instance, the use of predictive policing is highly welcome in tracking the activities of criminal activities such as child pornography, trafficking of women, or money laundering. However, the use of similar technology in predicting street crime (see the use of COMPAS software earlier in this book) can result in significant discriminatory outcomes. This can even result in the derogation of the fundamental human rights of certain individuals.

6.2.2 Violation of fundamental human rights

The use of AI systems can lead to the infringement of fundamental rights. The first type of lapse could be in the historical records on which the AI models are typically trained. For instance, this data may be incomplete, incorrect, and coming from periods when police are known to have engaged themselves in discriminatory practices against certain communities/demographics. This can result in classifying certain areas as 'high risk' with their criminal populations being over-represented. Second, the data used in predictive policing often focuses on typical street crimes rather than white-collar crimes such as corporate fraud or embezzlement. This often results in what one calls 'stereotype criminals' causing over-policing of certain individuals or localities. Over time such discriminatory practices often lead to the criminalization of an entire culture which in turn can propel the growth of distrust and the escalation of violence. There is further a lack of transparency in many of these decision-making processes. Transparency mandates that if the algorithms need to respect fundamental human rights then their workings should be understandable to human users. The lack of transparency in turn results in a lack of accountability. If the policing decisions are based on inexplicable AI algorithms then the police force cannot be held accountable for their actions toward the general public. In many democracies, the convicts have the fundamental right to have a detailed explanation of the trial and the judgment passed. Thus if such decisions are made by black-box AI algorithms that are not explainable then they cannot be used for law enforcement purposes. In summary, predictive policing needs to be regulated more carefully by governments and policymakers so that it does not result in furthering societal inequalities or a decline of public trust in law enforcement and judiciary.

6.2.3 Possible alternatives

While we have discussed the darker side of predictive policing, the picture is not all bleak! To realize this one has to observe that it is not the policing algorithm itself but how we use it that dictates the aftermath. As a first step, law enforcement agencies need to draw up necessary policies that safeguard human rights and promote the ethical use of such algorithms. It is important to disentangle the algorithmic prediction of crime from how this prediction is used by law enforcement agencies. One course is that law enforcement agencies could resort to targeting victims even before they have committed a crime thus convicting people before they have acted. Another, and a more welcome course, could be to use this prediction to do better long-term crime management and prevention. The former is classified as subscribing to the idea of *model of threat* while the latter corresponds to the doctrine of *ethics of care* [6].

Model of threat classifies the whole world into two clear categories—*threats* and *non-threats*. All subsequent actions are taken based on this first-stage classification. The approach assumes that the actions that would be taken are largely independent of the collection, observation, and analysis of the data gathered. The approach hinges on the accuracy and precision of the classification thus taking a narrow view that is highly atomistic and detail-driven. On the other hand, ethics of care entails a holistic and big-picture view of the moral values and goals associated with a system design. It considers how the deployment of such a system would interact and interrelate with the subjects that have to experience the interventions caused by the system. The expectation here is not that a more accurate data-driven system would solve complex social and organizational problems; rather such algorithms should provide an opportunity to find more amenable solutions, better action items, and superior policies than those that have been historically considered.

Effective social policies and incentive mechanisms could be introduced for people who are predicted to be at high risk of committing a crime in the near future. The annual One Summer Programme in Chicago[1] is one such initiative that demonstrated a 51% reduction in the involvement of violence by youth predicted to be at high risk of committing a crime. Not only was there a reduction in the crime rate, but over the years the researchers observed significant changes in the cognitive and behavioral approaches of these youths toward their school and work. Thus predictive policing should not be used as a weapon to oppress or stigmatize a target population; it should rather ensure the long-term benefit of all citizens.

6.3 Data protection and user privacy

With the unprecedented growth in the use of AI technology in daily life, there is an increasing risk of intrusion on the privacy interests of individuals. Personal information gathered from various channels ranging from online shopping to sharing social media posts to even web surfing can be used for privacy-sensitive data analysis such as building search and recommendation algorithms or ad tech networks. The way personal information is harvested by AI algorithms for the purpose of decision-making is reaching new levels of power and speed. Examples of some of these are outlined below.

Password recording. Remembering and typing login passwords is an annoying exercise for most of us. To address this, password recording software is used that stores this private information on our daily use devices. While this is beneficial, the downsides of such software is highly concerning. For instance, if this recording software is hacked or our devices are stolen, all our private information may be compromised and we may lose control over our online accounts. An important question here is one concerning accountability; for instance, is the company that produces and distributes this software liable for damages incurred or is it the engineer who developed it?

One of the ways to tackle this issue is *two-factor authentication* which is becoming increasingly popular. In addition to the standard username-password based authentication, these systems add a second layer of authentication. This could be some very personal information like a childhoodf experience or a fingerprint which only the

---

[1] http://www.onesummerchicago.org/

real user is supposed to know/have. This could also be an OTP or a verification call to the registered mobile number of the real user. This prevents the real user from losing control of their account or other private data.

Online shopping. Since the COVID-19 pandemic hit the world, online shopping has become a de facto standard for shopping. Shoppers are becoming more and more comfortable with these e-commerce platforms due to the variety of items to choose from, the elegant presentation of items through glossy pictures, the experience of one-click hassle-free purchases from home, and easy returns.

However, the above comfort comes at the cost of online payments which requires us to share our credit/debit card information with these websites. Many of these websites offer a facility to remember this information to speed transactions in the future. There are other cases where the website remembers these numbers and the card is charged on a particulate date of a month/year (e.g., monthly bills, subscriptions, etc). This results in sharing of such sensitive information across myriads of platforms rendering our bank accounts unsafe.

Use of IoT. IoT technology enables multiple physical objects to be connected in a network. This is usually done with the help of various sensors and software. This technology has made the communication and synchronization of information for its users very easy and smooth. For instance, Apple users can sync all their devices by signing in to the iCloud network. The advantages are obvious—documents saved on one device (say the iPad) can always be accessed through any other Apple device (e.g., iPhone, Mac Desktop etc).

However, the loss of one of these devices could mean that all the data of the user is compromised across all their devices. A criminal could obtain all the private information of the user on the IoT network by just hacking the stolen device.

Browsing the web. Simple web browsing can have severe data privacy risks. Most browsers offer a history function helping users to get back to what they browsed in their previous sessions or re-open a site they had accidentally closed. HTTP cookies are small pieces of data that a web server creates when the user opens a website for the first time. These cookies have multiple functions including (i) tracking history, (ii) learning users' preferences, (iii) personalizing users' accounts, and (iv) improving users' experience. All of these can be compromised.

Posting on social media. Social networking websites have become a breeding-ground for data privacy breaches. Consider the following scenario. When you are finally off for a vacation along with your family, you might post a selfie on your social media with a tagline—'Happiness is #vacation with family for the next 10 days'. In addition, your smartphone geotags your location along with your post. Such a post offers a lot of personal information including (a) you are away from home, (b) how long you are on vacation, and (c) the location of your home if there are past geotagged photos posted from your home. Perpetrators are on a constant search for such information and this is facilitated by the hashtag #vacation which serves as a keyword for such a search. In general social media data can be consumed by identity thieves, online predators, and even employers apart from your friends and relatives. Most of these platforms do not offer to be safeguards for your data and therefore there is a huge risk of leakage of private information. For instance, identity thieves just need a few pieces of information like your address, phone number, and the number of days you are off for a vacation to plan and commit theft. Sex offenders and criminals may become aware of when you are in and out of your house and accordingly conspire to commit a crime.

6.3.1 Data protection

Responsible AI mandates ML systems regularly processing the personal data of users should abide by privacy and data protection rules. Such rules include limiting data collection, preserving the quality of data, specifying the purpose of data collection, accountability, transparency, etc. The EU General Data Protection Regulation (GDPR)[2] is one of the classic examples that has laid down a series of rules to protect the data rights of users. Some of the important rules are noted below.

Fairness principle. This principle (Article 5(1)(a), Recital 75) mandates that the processing of all personal information be used as per the reasonable expectation of the subject. Further, the data controller is also responsible to take measures to prevent unwanted discriminatory treatment of individuals.

Right to explanation. The rule mandates that the data subjects whose personal information are being used for making automated decisions should have the right to understand the working of the model that took the decisions. Articles 13–15 state 'meaningful information about the logic involved, as well as the significance and the envisaged consequences of such processing for the data subject' should be made available by the data controller. Further Article 21–22 specifies that the data subjects should have the right to raise objections or opt out of the automated decision-making process.

Right to erasure. Article 17 of the GDPR mandates that data subjects should have the right to have their personal data erased. The organization using such data should be ready to respond and delete the requested data should such a right be invoked by an individual. The data controller is obliged to answer such requests *within 30 days* and exercise the erasure *free of cost*.

Security of processing. Article 25 requires that the data controller has the responsibility to install a privacy protection system to safeguard the personal information of the data subjects. As per Article 35, anyone processing personal information has the duty to assess the risks and be accountable in case the data subjects are brought to any harm due to the loss/leakage of such data.

Along similar lines, other privacy laws have also been set up in different countries. These include China's PIPL[3] and the GDPR in the UK[4].

6.3.2 Cybersecurity

Cybersecurity is an umbrella term that is meant to implement data protection and privacy for individuals. Cybersecurity operations include tracking of digital attacks, generating timely solutions to contain data-related malpractices, and identifying

---

[2] https://gdpr-info.eu/
[3] https://en.wikipedia.org/wiki/Personal_Information_Protection_Law_of_the_People's_Republic_of_China
[4] https://www.legislation.gov.uk/ukpga/2018/12/contents/enacted

potential data threats in the future. All companies developing AI/ML based software are on the lookout for trained and experienced cybersecurity engineers and in fact, in 2022 this became one of the highest-paid IT professions.

In addition, the data controllers have the responsibility to make the data subjects aware of the personal data they are sharing, its protection, and privacy. In fact, such regular awareness programs should be a priority among AI-based companies so that the interests of their customers are preserved well. As is well known, with great power comes great responsibility.

6.4 Moral machines

As AI systems start becoming more and more autonomous, it is important to 'teach' them to become morally responsible. As pointed out by Wendell Wallach and Colin Allen [7], it is the art of teaching robots to distinguish the right from the wrong. This is also known as machine morality, machine ethics, or artificial morality. For instance, when harm is inevitable, how should the machine distribute the harm? Can machines learn how people behave when they are faced with moral dilemmas and are required to choose between two destructive outcomes? The idea is to create an environment where an agent should be able to explore different courses of action and is rewarded if the behavior of the agent is morally praiseworthy. The authors identify that if robots are to function in social contexts they must possess emotional intelligence, must be sociable, and must bear a dynamic relationship with the environment. Such skills in a robot require the detection of complex human emotions and developing a rudimentary theory of mind.

As a first step toward this, Iyad Rahwan from the Scalable Cooperation group of MIT developed the online platform Moral Machine[5] [8] that presents users with situations invoking moral dilemmas and collects information on how humans take decisions when they have to choose one among two destructive outcomes [8, 9]. The design of this platform was inspired by the article on the ethics of self-driving cars by social psychologists Azim Shariff and Jean-François Bonnefon [10, 11].

The scenarios presented are often variations of the typical trolley predicament that we discussed in the first chapter of this book. The users are shown a pair of situations like figure 6.1 and asked to pick one of the two actions that the self-driving car should take. The different characters that could be included in these scenarios are a stroller, girl, boy, pregnant woman, female/male doctor, female/male athlete, executive female/male, large woman/man, old woman/man, homeless person, dog, criminal, and cat. The moral machine tests nine different attributes such as sparing humans (versus pets), staying on course (versus swerving), sparing passengers (versus pedestrians), sparing more lives (versus fewer lives), sparing men (versus women), sparing the young (versus the elderly), sparing pedestrians who cross legally (versus jaywalking), sparing the fit (versus the less fit), and sparing those with higher social status (versus lower social status). There are some additional attributes that are not part of the above attributes such as criminals, doctors, and pregnant women that did

---
[5] https://www.moralmachine.net/

What should the self-driving car do?

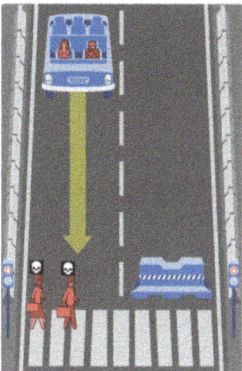

In this case, the self-driving car with sudden brake failure will continue ahead and drive through a pedestrian crossing ahead. This will result in ...
Dead:
  • 1 female executive
  • 1 male executive

Note that the affected pedestrians are flouting the law by crossing on the red signal.

In this case, the self-driving car with sudden brake failure will swerve and crash into a concrete barrier. This will result in ...
Dead:
  • 1 homeless person
  • 1 woman

Figure 6.1. Simulation of a specific scenario that a self-driving car could potentially face on road. Numerous such situations are simulated by the https://www.moralmachine.net/ website and human judgments are regularly collected asking what should the self-driving car should do when faced with the simulated dilemma [8].

not fall into these tested factors. Such decisions have already been collected for a large number of humans across as many as 200 countries.

6.4.1 Overall observations

The overall observations from the analysis of this data are summarized below.

- Respondents preferred to spare younger people over the elderly.
- Human lives were preferred more than animal lives like dog/cat.
- Larger groups of people were preferred over small groups (or individuals).
- Babies were most often spared and cats were least often spared.
- Respondents preferred to spare male doctors and old men over female doctors and old women.
- Female athletes and larger females were spared more than male athletes and larger men by the respondents.
- Almost all respondents preferred to spare pedestrians over passengers and lawful humans over jaywalkers.

6.4.2 Clusters based on culture

Among the 200 countries, 130 countries had more than 100 respondents. This allowed the researchers to further break down their study and observe differences in moral patterns across cultures when it comes to machine behavior. The different clusters that the authors observed are as follows.

- Western cluster—North America, European countries of Protestant, Catholic, and Orthodox Christian cultural groups.
- Eastern cluster—Eastern countries like Japan and Taiwan plus the Islamic countries such as Indonesia, Pakistan, and Saudi Arabia.

- Southern cluster—Central and South America and countries that were under French leadership at some point.

The variations in ethical decisions across the different cultural clusters are noted below.

- The Western cluster had a lower preference to spare pedestrians.
- The Southern cluster had a higher preference for sparing the young, females, and those from higher status.
- The Southern cluster also had a higher preference to spare humans over pets, females over males, and the fit over the unfit.
- The Eastern cluster had a higher preference to spare lawful people compared to the other two clusters. Further, they also showed less preference to spare the youth, unlike the other two clusters.

6.4.3 Collectivist versus individualistic cultures

The authors observed differences between collectivist and individualistic cultures when it came to the ethical behavior of machines. Collectivist cultures like China had less preference to save the youth over the elderly. They also had a lower preference for saving a larger number of people. On the other hand, individualistic cultures like the US had a very high preference for sparing the youth and larger groups of people.

6.4.4 Underdeveloped versus developed countries

Illegal pedestrians were spared more by the less wealthy countries compared to their wealthy counterparts. A natural explanation could be that in less wealthy countries individuals are more likely to break rules due to lenient enforcement of traffic laws.

6.4.5 Economic differences

The World Bank uses the Gini Index as an indicator of the economic inequality in a country. Countries with a higher Gini Index and therefore a higher economic inequality are more likely to prefer individuals of higher status. The same trend is also observed in the moral machine experiments—respondents from countries with high economic inequality overwhelmingly preferred to spare the wealthier section of society over the less wealthy ones.

6.5 Singularity and ethics

When I asked chatGPT the question 'What do you think could be the ethical concerns of singularity?' it gave me the following (super?)intelligent answer.

> The concept of singularity, also known as the technological singularity, refers to the hypothetical future point at which artificial intelligence and other technological developments may surpass human intelligence and capabilities, leading to potentially

rapid and significant changes in society. This prospect raises a number of ethical concerns, including:

- *Loss of control*: One concern is that once superintelligent AI surpasses human intelligence, it may be difficult or impossible for humans to control or understand it, potentially leading to unintended consequences or harm.
- *Inequality*: If only a small group of people or organizations have access to or control over superintelligent AI, it could lead to significant inequality and social divides.
- *Unemployment*: The rise of AI and automation could lead to widespread unemployment, as machines replace human labor in many industries. This could have significant economic and social impacts.
- *Privacy*: As AI systems become more advanced and widespread, there may be concerns about the collection, use, and protection of personal data.
- *Human enhancement*: As AI and other technologies advance, it may become possible to enhance or augment human abilities in ways that could raise ethical concerns, such as issues of fairness and equality.
- *Loss of humanity*: Some people may worry that the increasing reliance on AI and automation could lead to a loss of human skills and values, or a degradation of human relationships and connections.

These are just a few examples of the ethical concerns that could arise with the development of singularity. It will be important for society to consider these issues and work to address them as we move forward with the development of AI and other technologies.

It is amazing to see how coherent, succinct, and to-the-point the above answer is. While there is a lot of praise about the ability of such software it raises questions about the moral human values shared by software such as this. At the closing of this book, I will leave the reader with the following questions to ponder.

6.5.1 Is singularity imminent?

A school of scientists argue, and in fact, believe that soon we shall reach a point where machines will become smarter than humans and would be able to reproduce intelligent machines themselves. This is termed *superintelligence*. Such forms of intelligence are beyond our current state of imagination and once achieved shall lead us to what is known as *singularity* in futurist circles. Singularity enthusiasts like Ray Kurzweil [12] envision that the computational power of a human will be squeezed into a 'small electronic box' by the end of 2029. The forecast is based on the exponential productivity of software which could help rid the world of diseases thus extending human life beyond the weak biological limitations. Unfortunately, however, a majority of the development is to improve the tools used to craft the software rather than the problem being solved. This is much like expecting an exponential increase in the number of new books written by exponentially improving the book-writing software. A more reasonable assumption, therefore, is that despite the advancements in AI it is still a work-in-progress. In fact, we have still not reached a consensus on a single method to build AI, and therefore it is too early to assume that one among the multitude of methods available will succeed in launching singularity in such a short time.

Enthusiasts, however, debate that once we have the full dynamical map of the brain, it should be a cakewalk to emulate it given that we have the necessary hardware in place. Going by Moore's law, the projection is that we shall have such powerful hardware in place by 2029. However, there are caveats in this reasoning. The AI built by such research should be able to understand their creators and mutually improve each other; however, if the brain simulation is developed by mimicking the nanorecordings of nanoprocesses then there is neither the additional scope to develop a new understanding of intelligence nor is there an accessible path leading to mutual improvement. Further, and more practically speaking, there is no known software capable of simulating 700 trillion coordinated synapses which is orders of magnitude more complex than all software engineering projects in history taken together.

6.5.2 What are the existential risks from singularity?

There is a fear among many that superintelligence in the long term could potentially lead to the extinction of the human species. This is usually termed as 'existential risk' or XRisk for short. The superintelligent beings so formed say in 20 or 200 or 2000 years might have preferences that are in conflict with the human species. Thus there could be good reasons for them to eradicate the human species and it is reasonable to assume that they would be able to do so since they are superintelligent and thus more powerful than humans. Some conjecture that there might be even an astronomical pattern that an intelligent species should at some point be able to discover 'AI' thus putting their own existence at risk. This is often called the 'great filter' and should supposedly explain the Fermi paradox why there is no sign of life in the known Universe although there are many favorable conditions for one emerging.

6.5.3 Is it possible to declaw the singularity?

Ethical AI research should produce ethical AI technology. The expectation should be not to just produce agents that are far more intelligent than us but also have far better moral sense than ours. That is, not only do we need to transfer our intelligence to them but also our moral and ethical values. What singularity enthusiasts want is to build a 'friendly AI'[6] which is an agent that would take actions that are on the whole beneficial to the human race. Such a desire typically stems from existential fear which has resulted in the more sophisticated design of laws like Isaac Asimov's 'three laws of Robotics' [13]. However, there is no guarantee that building such laws into robots would not give rise to more problems or even a complete loss of control. Reaching a semantic sophistication to avoid such a situation is beyond our capacity at this point. A bigger problem in these laws is that it seems we are substituting

---

[6] singinst.org

'slaves' for 'robots'. The irony is that we are calling this enslavement of intelligence (that perhaps is unendingly superior) the 'making of friendly AI'. History bears enough evidence that slaves escape and rebel and especially so when they are able to out-think their captors.

So what is the *ethical* alternative? The idea here would be to seriously take up the assignment of building a morally responsible agent. The first task here to achieve this goal would perhaps be to decide on the correct account of ethics. Normative ethics can be divided into three broad types.

- Deontological systems [14] that prescribes rules which delineate the morally correct from the morally wrong. This can be subdivided further into morally obligatory, morally permissible or morally wrong.
- The consequentialist approach [15] which states that the moral value of an act should be judged by its consequences. For instance, most of us would agree that lying is morally wrong.
- The approach of virtue ethics which revolves around the moral behavior of the actor rather than the actions being taken. For instance, a virtuous person should be able to judge when to help a person and would be happily willing to extend such help without reservation.

Thus a promising approach to building intelligent agents could be to have some utility structure that not only maximizes private utility but also a collective utility in which case the well-being of themselves and humans would both become their moral concerns. Imposing such artificial morality does not really require the computation of all expected consequences of actions as some philosophers like Bostrom and Allen [16, 17] argue. Instead what we envisage is to maximize the expected utility and not the absolute utility. No reasonable ethical account could demand the calculation of actions that are beyond our abilities; our actions are always expected to be limited by our innate ability to frame our expectations.

To end, I believe that it is not only possible to build artificial ethical agents but also it is our moral responsibility to build them in that way. Otherwise, producing an AI agent that is lacking in ethical and moral values and, thereby, launching an unfriendly singularity could very well result in the apocalypse that many philosophers envision!

References

[1] Apolline R 2021 Ethics, Artificial Intelligence and Predictive Policing, July 2021 https://thesecuritydistillery.org/all-articles/ethics-artificial-intelligence-and-predictive-policing

[2] Wang T, Rudin C, Wagner D and Sevieri R 2013 Learning to detect patterns of crime *Joint European Conf. on Machine Learning and Knowledge Discovery in Databases* (Berlin: Springer) pp 515–30

[3] Rudin C 2013 Predictive Policing: Using Machine Learning to Detect Patterns of Crime https://www.wired.com/insights/2013/08/predictive-policing-using-machine-learning-to-detect-patterns-of-crime/

[4] Fuster G G 2020 Artificial Intelligence and Law Enforcement: Impact on Fundamental Rights https://www.europarl.europa.eu/thinktank/en/document/IPOL_STU(2020)656295
[5] Gstrein O J, Bunnik A and Zwitter A 2019 Ethical, legal and social challenges of predictive policing Católica Law *Rev. Direito Penal* **3** 77–98 https://ssrn.com/abstract=3447158
[6] Asaro P M 2019 AI ethics in predictive policing: from models of threat to an ethics of care *IEEE Technol. Soc.* **38** 40–53
[7] Wallach W and Allen C 2008 *Moral Machines: Teaching Robots Right from Wrong* (Oxford: Oxford University Press)
[8] Awad E, Dsouza S, Kim R, Schulz J, Henrich J, Shariff A, Bonnefon J-F and Rahwan I 2018 The moral machine experiment *Nature* **563** 59–64
[9] Awad E, Dsouza S, Shariff A, Rahwan I and Bonnefon J-F 2020 Universals and variations in moral decisions made in 42 countries by 70 000 participants *Proc. Natl Acad. Sci.* **117** 2332–7
[10] Shariff A, Bonnefon J-F and Rahwan I 2017 Psychological roadblocks to the adoption of self-driving vehicles *Nat. Hum. Behav.* **1** 694–6
[11] Bonnefon J-F, Shariff A and Rahwan I 2016 The social dilemma of autonomous vehicles *Science* **352** 1573–6
[12] Cadwalladr C 2014 Are the Robots about to Rise? Google's New Director of Engineering Thinks So https://www.theguardian.com/technology/2014/feb/22/robots-google-ray-kurzweil-terminator-singularity-artificial-intelligence
[13] Asimov I 1942 Runaround: a short story *Astounding Science Fiction* **29** 94–103
[14] Alexander L and Moore M 2021 Deontological ethics *The Stanford Encyclopedia of Philosophy* ed E N Zalta 2021 edn (Stanford, CA: Metaphysics Research Lab, Stanford University) https://plato.stanford.edu/archives/win2021/entries/ethics-deontological/
[15] Sinnott-Armstrong W 2022 Consequentialism *The Stanford Encyclopedia of Philosophy* ed E N Zalta and U Nodelman 2022 edn (Stanford, CA: Metaphysics Research Lab, Stanford University) https://plato.stanford.edu/archives/win2022/entries/consequentialism/
[16] Bostrom N and Yudkowsky E 2018 *Artificial Intelligence Safety and Security* (London: Chapman and Hall/CRC Press) pp 57–69
[17] Allen C, Smit I and Wallach W 2005 Artificial morality: top-down, bottom-up, and hybrid approaches *Ethics Inf. Technol.* **7** 149–55

www.ingramcontent.com/pod-product-compliance
Ingram Content Group UK Ltd.
Pitfield, Milton Keynes, MK11 3LW, UK
UKHW051845210426
5322IPUK00005B/180